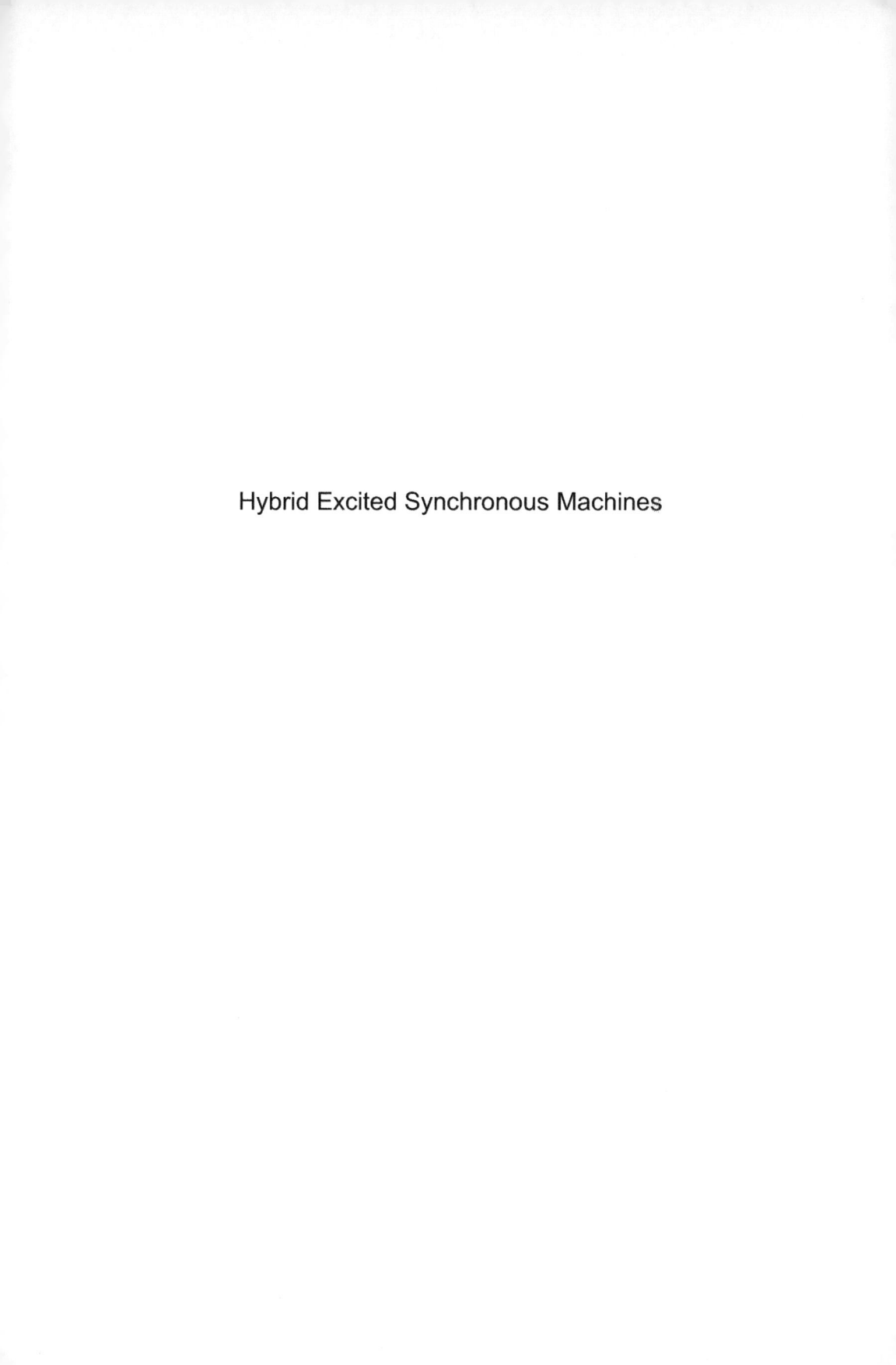

Hybrid Excited Synchronous Machines

*To my colleagues, confirmed and young researchers, who have contributed directly
or indirectly to the research effort on hybrid excited synchronous machines.
To my family on both sides of the Mediterranean Sea.
To my parents, those book lovers.
To my sister and brother and their relatives.
To my wife and kids.
Thank you!*
Yacine AMARA

*To the professors who have transmitted their knowledge and passion
for the design of non-conventional electrical machines:
Jean Lucidarme, Mohamed El Hadi Zaim, Bernard Multon,
Christian Rioux, Rachid Ibtiouen
and many others in France and elsewhere.*
Hamid BEN AHMED

*I would like to dedicate this book to the entire team
who worked with me on this fascinating subject.
A special mention to Michel Lécrivain, Emmanuel Hoang,
Sami Hlioui, Franck Chabot and Cédric Plasse.*
Mohamed GABSI

Hybrid Excited Synchronous Machines

Topologies, Design and Analysis

Yacine Amara
Hamid Ben Ahmed
Mohamed Gabsi

iSTE   WILEY

First published 2022 in Great Britain and the United States by ISTE Ltd and John Wiley & Sons, Inc.

Apart from any fair dealing for the purposes of research or private study, or criticism or review, as permitted under the Copyright, Designs and Patents Act 1988, this publication may only be reproduced, stored or transmitted, in any form or by any means, with the prior permission in writing of the publishers, or in the case of reprographic reproduction in accordance with the terms and licenses issued by the CLA. Enquiries concerning reproduction outside these terms should be sent to the publishers at the undermentioned address:

ISTE Ltd
27-37 St George's Road
London SW19 4EU
UK

www.iste.co.uk

John Wiley & Sons, Inc.
111 River Street
Hoboken, NJ 07030
USA

www.wiley.com

© ISTE Ltd 2022

The rights of Yacine Amara, Hamid Ben Ahmed and Mohamed Gabsi to be identified as the authors of this work have been asserted by them in accordance with the Copyright, Designs and Patents Act 1988.

Any opinions, findings, and conclusions or recommendations expressed in this material are those of the author(s), contributor(s) or editor(s) and do not necessarily reflect the views of ISTE Group.

Library of Congress Control Number: 2022943360

British Library Cataloguing-in-Publication Data
A CIP record for this book is available from the British Library
ISBN 978-1-78630-685-2

Contents

Foreword vii

Introduction ix

Chapter 1. Hybrid Excited Synchronous Machines: Principles and Structures 1

    1.1. Introduction. 1
    1.2. Interest in hybrid excitation 3
        1.2.1. Motoring mode operation 4
        1.2.2. Generation mode operation 10
    1.3. Hybrid excited structures. 12
        1.3.1. Classification criteria. 13
        1.3.2. Structures and classification 19
    1.4. Conclusions and perspectives 28

Chapter 2. Control of Hybrid Excited Synchronous Machines 31

    2.1. Introduction. 31
    2.2. Modeling of hybrid excited synchronous machines 32
        2.2.1. The nature of the equations 35
        2.2.2. Control modes. 39
    2.3. Torque characteristics and basic control laws. 41
        2.3.1. Torque characteristics as a function of I and ψ 42
        2.3.2. Torque characteristics as a function of V and δ 44
        2.3.3. Notion of stability for an open loop and the consequences of closed-loop operations. 45
        2.3.4. Fundamental control laws 51
        2.3.5. Temporary overloaded motor operation. 56
    2.4. Setting the speed of HESMs (maximal characteristics/envelopes) . . . 58

2.4.1. Low-speed operations 59
2.4.2. Operation at high speeds/the notion of flux weakening 72
2.5. Operations on the entire "torque/speed" plane 111
2.5.1. Efficiency optimization algorithms on the entire
"torque/speed" plane. 113
2.5.2. Normalized model with losses and the calculation of $V_{n\,max}$. . . . 118
2.5.3. Machines with non-salient poles ($\rho = 1$) 121
2.5.4. Machines with salient poles ($\rho \neq 1$) 125
2.5.5. Validity of the tools developed and the contribution towards
hybrid excitation 131
2.6. Conclusions and perspectives 148

Chapter 3. Experimental Studies of Hybrid Excited
Synchronous Machines 153

3.1. Introduction. 153
3.2. Machine 1. 154
3.2.1. Structure and operating principles 155
3.2.2. Construction 159
3.2.3. Experimental study. 162
3.3. Machine 2. 170
3.3.1. Structure and operating principle. 172
3.3.2. Construction 179
3.3.3. Experimental study. 186
3.4. Conclusions and perspectives 194

Conclusion. 197

References. 199

Index 211

Foreword

Hybrid excited synchronous machines consist of two excitation sources, i.e. permanent magnets and field windings. Thus, they may have the synergies of both permanent magnet machines and wound field machines, most notably the high efficiency and high torque density of permanent magnet machines, and the flexible flux adjustment of wound field machines. There are many machine topologies of hybrid excited synchronous machines according to the allocation of permanent magnet and field winding excitations, series and parallel magnetic circuit hybridization, as well as two- and three-dimensional hybrid magnetic circuits. The introduction of field windings can help regulate the magnetic field and the output capability according to the working condition, which is beneficial for many applications, such as electric vehicles, with variable-speed requirements. Compared with the conventional permanent magnet machines, an extra flexibility can be used to adjust the flux linkage by the field excitation current. Consequently, higher torque at low speed, wider operating speed range, as well as high efficiency over a wide operating region can be obtained by employing appropriate control strategies. Hybrid excitation also provides a unique mitigation technique for an uncontrolled overvoltage fault at high-speed operation.

The authors of this book have been working on hybrid excited synchronous machines and control systems for many years. They have led the development of several novel hybrid excited synchronous machine topologies. The book is comprehensive in that it includes not only hybrid excited synchronous machines but also their control strategies. I am also very pleased to see the experimental investigation. Currently, hybrid excited synchronous machines still exhibit some challenges, such as complicated structures and relatively low torque density. I am sure that this book will serve researchers, students and engineers very well in their further investigation for various applications.

Professor Zi-Qiang ZHU,
Fellow of Royal Academy of Engineering,
Fellow of IEEE, Fellow of IET
University of Sheffield, UK

Introduction

The increasing use of electrical energy in the functioning of our society, which is related to our transition towards clean and sustainable uses of energy, requires tools and methods to study electrotechnical objects to be developed to ensure their proper functioning for as long as possible. This means that we must have the tools required to understand their behavior, not just as one component, but also in the entire systems in which they can be found and throughout their life cycle, and therefore under any operating conditions.

As electromechanical conversion devices, electrical machines are at the heart of electrotechnical systems and energy conversion chains. At one extreme, they are used as a source of electrical energy (electrical generators), and at the other extreme, they are used to transform electrical energy into mechanical work (electric motors and actuators). These functions are performed very efficiently by such machines.

The purpose of this book is to provide electrical engineering students and researchers with the tools to analyze how synchronous machines operate over their entire field of operation.

In this book, particular emphasis is placed on these hybrid excited synchronous machines (HESMs). This HESM theme, although not a fundamental problem in the strict sense of the term, provides answers to problems which are no less fundamental: the flux weakening of permanent magnet machines, energy optimization and finally the increasing costs of rare-earth permanent magnets.

Among all the various types of electrical machines, permanent magnet synchronous machines are the most efficient when their operation does not require any flux weakening. This is due to the permanent magnets which produce an excitation flux with next to no losses. However, this powerful flux created by the magnets is a disadvantage when operating at high speeds because flux weakening is required.

From a functional point of view, HESMs combine the advantages of permanent magnet machines (which have very high energy efficiencies) with those of electrically excited machines (which have an ease of operation along with variable speeds). In these machines, the total excitation flux is the sum of the flux created by the permanent magnets and the excitation flux created by the wound coils. This hybridization allows for greater flexibility for variable-speed operation and, in parallel, energy efficiency optimization for certain drives.

The authors have been working on HESMs for more than two decades, and this book is both a collection of their work and an update of the work done on drives based on synchronous machines (Lajoie-Mazenc and Viarouge 1991; Amara et al. 2009; Vido et al. 2011). It may be useful for both experienced researchers and students who wish to deepen their knowledge of synchronous machines.

The study of HESMs allows us to understand the behaviors of other families of synchronous machines. Permanent magnet synchronous machines can be considered as a special case of HESMs, where excitation coils are not used; electrically excited machines correspond to HESMs in which magnets are replaced by air; and finally, variable-reluctance synchronous machines correspond to HESMs in which the magnets are replaced by air, and wound field excitation is not used.

This book begins with a presentation of the principle of hybrid excitation for synchronous machines. The first chapter sets out the concept of what is meant by an HESM, placing them within the vast family of electrical machines. An up-to-date state of the art is presented, where the focus is on the HESM component. Different structures encountered in the scientific and technical literature are analyzed.

The second chapter is devoted to the behavior of HESMs when they operate as a motor. An original study of the behavior of these machines and their environment is presented. It allows us to analyze the operation of HESMs, by taking their power supply into account (De Doncker et al. 2011). Electrical circuit models are used for this purpose. The purely analytical approach will be favored to determine the control quantities and the different control laws, according to the parameter sets used for the models. When the analytical approaches become too complicated to develop further, scripts that allow us to continue the analysis numerically will be made available to the reader. The behavior of the HESMs is analyzed over their entire operating range.

This book ends with a presentation of some structures that have been designed, realized and studied in our respective laboratories (SATIE and GREAH). This chapter illustrates the previous concepts and theoretical developments concretely.

We hope that this book will contribute to a better understanding of how HESMs function, and to the awakening of new ideas within our research laboratories.

1

Hybrid Excited Synchronous Machines: Principles and Structures

1.1. Introduction

Hybrid excited machines form a special class within that of synchronous machines. Hybrid excited synchronous machines (HESMs) are those in which two sources of excitation flux coexist (Figure 1.1): permanent magnets and excitation coils. Different terminologies (we will limit ourselves to terms in both English and French) have been used to describe this type of machine:

– hybrid excited (or excitation) synchronous machines (*Machine synchrones à excitation hybride*);

– double excitation synchronous machines or dual excitation synchronous machines (*Machine synchrones à double excitation*);

– combined excitation synchronous machines (*Machines synchrones à excitation combinée*);

– permanent magnet synchronous machines with auxiliary excitation windings (*Machines synchrones à aimants permanents avec des bobinages d'excitation auxiliaires*).

The aim of this combination is to combine the advantages of permanent magnet machines with those of electrically excited synchronous machines (Spooner et al. 1989; Levy 1997; Mizuno 1997; Hadji-Minaglou and

---

For a color version of all the figures in this chapter, see www.iste.co.uk/amara/hybrid.zip.

Henneberger 1999; Luo and Lipo 2000; Amara 2001). Figure 1.1 illustrates the principle of hybrid excitation, where the total excitation flux is the total sum of the contributions from the permanent magnets and the excitation winding.

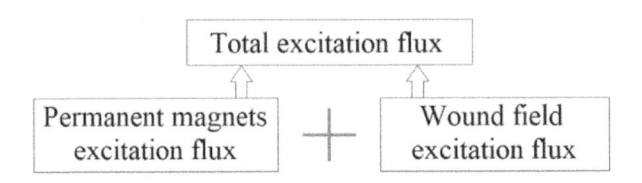

Figure 1.1. *Illustration of the hybrid excitation principle*

The effective performance of these machines, in terms of operational flexibility at high speeds and their improved efficiency, has received a growing interest from researchers in both the academic and industrial worlds (Syverson 1995; Amara et al. 2004a, 2009; Akemakou 2006; Aydin et al. 2007; Mizutani et al. 2008; Reutlinger 2008; Wang 2009; Yu et al. 2009; Fang et al. 2010; Moynot et al. 2010; Wang et al. 2010; Xia et al. 2010; Xie and Xu 2010; Greif et al. 2011; Hoang et al. 2011; Pothi et al. 2015; Chao et al. 2018; Diao 2018; Liu et al. 2018).

In addition to this positive contribution towards synchronous machine operation, hybrid excitation also allows for flexibility in their design. Indeed, the adjustment of the two excitation sources constitutes an additional degree of freedom in this respect. This degree of freedom is formalized by a parameter, which we will call the hybridization rate α, and is defined by:

$$\alpha = \frac{\Phi_a}{\Phi_{exc\,max}} \qquad [1.1]$$

where $\Phi_{e\,max}$ is the maximum total excitation flux, and Φ_{PM} is the permanent magnet flux linkage.

In motoring mode, these machines provide greater flexibility at high speeds and optimized energy efficiency (Amara 2001; Amara et al. 2004a, 2009; Pothi et al. 2015). In generating mode, hybrid excited synchronous machines that are connected to a diode rectifier are an interesting alternative to permanent magnet machines associated with a controlled rectifier for high-speed operations (Amara et al. 2007). Moreover, in the current state of increasing costs for high-performance permanent magnets (rare-earth materials), as well as the negative ecological impact when extracting these rare-earth materials, the principle of hybrid excitation reduces the volume required for magnets in synchronous machines.

1.2. Interest in hybrid excitation

The work by Spooner et al. (1989) can be considered as one of the earliest academic contributions towards the topic of hybrid excited synchronous machines. However, it should be noted that patent applications were made much earlier (Rosenberg 1940; Shafranek 1967; Rosenberg 1968).

In the patent application (Shafranek 1967), the main claim concerned the provision of hybrid excitation to ensure the self-excitation of the alternators. Figure 1.2 shows the machine that was the subject of the patent. It should be noted that every source of magnetic flux, permanent magnets, excitation windings, as well as armature windings, are situated within the stator. It should be noted that not every structure illustrated in this chapter necessarily comes from a design.

The patent application by Rosenberg (1968), as well as the contribution from Spooner et al. (1989), discuss precisely what is meant by the principle of hybrid excitation as treated in this monograph. The authors express an interest in hybrid excitation as a means of controlling the air-gap flux. Moreover, the Spooner et al. (1989) authors point out the economic interest in reducing the volume used for permanent magnets.

In these previous two references (Rosenberg 1968 and Spooner et al. 1989), the authors emphasize the interest in controlling the air-gap flux obtained by the means of hybrid excitation, both for the motoring and the generating modes.

The contribution coming from hybrid excitation is illustrated in the following for both operational modes; in particular, the contribution from a functional viewpoint is emphasized.

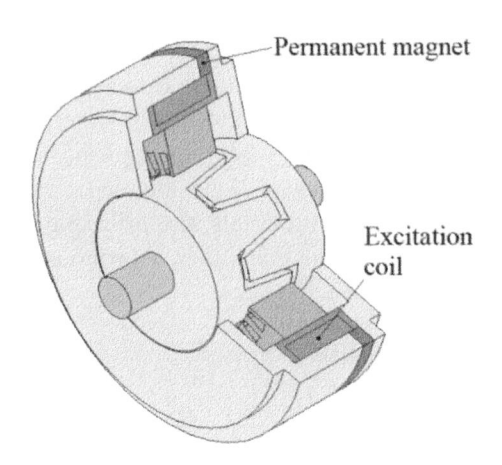

Figure 1.2. *Self-excited brushless alternator (Shafranek 1967)*

1.2.1. *Motoring mode operation*

For motoring operational modes, the contribution from hybrid excitation is two-fold (Amara et al. 2019). First, it improves the flux weakening capability by controlling the total excitation flux. This allows for a broader operational range while maintaining a constant power, even for machines with a relatively low armature magnetic reaction field when compared to the excitation flux. A direct consequence of this is the possibility for the machine to operate with an effective power factor, which allows for a better dimensioning of the converter-machine assembly. Second, it improves the energy efficiency of the drive train.

To illustrate the first contribution, a simple theoretical development is presented in what follows: the framework for this development is the application of electric traction. It is based on the simplified hybrid excited machine model with smooth poles (no saliency), for which we ignore any losses and any magnetic saturation. We will return to this operational mode in more detail in Chapter 2.

Figure 1.3 shows the equivalent electrical diagrams in the d- and q-axes of the "Park" frame of the first harmonic model with no loss, where:

– i_d, i_q: d- and q-axis components of the armature current (A);

– v_d, v_q: d- and q-axis components of the terminal voltage (V);

- L: synchronous armature inductance (H);
- Φ_a: permanent magnet flux (Wb);
- Φ_b: wound field excitation flux (Wb);
- Φ_{exc}: total excitation flux (Wb);
- ω: electrical pulse (rad/s).

Figure 1.4 shows the vector diagram representing the main electromagnetic quantities (flux, EMF, voltage, current) defined in the "Park" reference frame. By convention, the angles ψ and δ, as they appear in this figure, are counted negatively and positively, respectively.

The armature flux linkage Φ_{arm} is the vector sum of the excitation flux Φ_{exc}, and the flux of the armature magnetic reaction Φ_{AMR}:

$$\vec{\Phi}_{arm} = \vec{\Phi}_{exc} + \vec{\Phi}_{AMR} \qquad [1.2]$$

Despite being scalar quantities, it is possible to associate a vector to each flux, which in the framework of the first harmonic hypothesis is carried out using the axis corresponding to the spatial position at which each flux attains its maximum value. The excitation flux amplitude is given by:

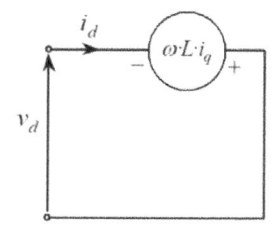

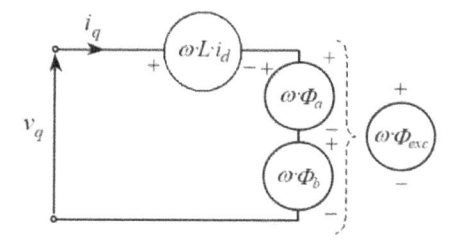

$$\Phi_{exc} = \Phi_a + \Phi_b = k_f \cdot \Phi_{exc\,max} \qquad [1.3]$$

a) d-axis equivalent electrical circuit.

b) q-axis equivalent electrical circuit.

Figure 1.3. *Equivalent electrical circuits for non-salient hybrid excited synchronous machines (lossless model)*

Having defined the two components for the armature flux linkage [1.2], it is interesting to distinguish between two terms that are often found in the literature: flux weakening and "de-excitation". The first refers to the reduction of the total armature flux linkage, and the second to the reduction of the excitation flux. A flux weakening strategy may require de-excitation, and we will see in Chapter 2 that optimal flux weakening strategies do indeed often require de-excitation. The term "optimal" is used to describe any strategy that maximizes the torque or the power during high-speed operations.

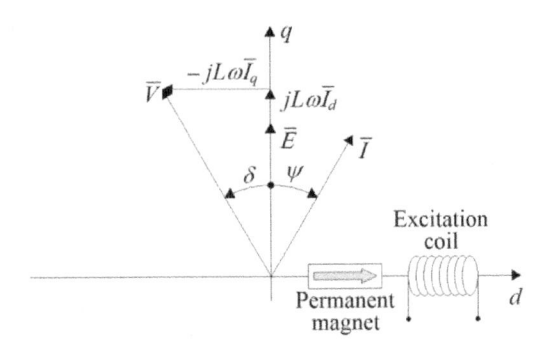

Figure 1.4. *Equivalent vector diagram for the lossless model of non-salient HESMs (motoring mode)*

In this model, the modulus of the armature voltage and the value of the electromagnetic torque are, respectively, given by:

$$V = \Phi_{arm} \cdot \omega = \Phi_{arm} \cdot p \cdot \Omega \qquad [1.4]$$

$$\Gamma_{em} = p \cdot \Phi_{exc} \cdot i_q = p \cdot \Phi_{exc} \cdot I \cdot \cos(\psi) \qquad [1.5]$$

Equation [1.4] relates the armature voltage modulus V to that of the armature flux linkage Φ_{arm} and to the angular speed Ω. Equation [1.5] relates the electromagnetic torque Γ_{em} to the excitation flux modulus Φ_{exc} and to that of the armature current I. Here, p is the number of pole pairs.

Figure 1.5 shows the characteristic envelopes of electrical drive specifications. Figure 1.5(a) shows the voltage, or power, characteristics as a function of the speed, and Figure 1.5(b) shows the torque characteristic.

The power has a characteristic whose shape is identical to that of the voltage. Below a certain speed, called the base speed Ω_b, the torque must be as high as possible (Figure 1.5(b)). To impose the maximum amount of torque, the modulus of the excitation flux and the armature current, labeled $\Phi_{exc\ max}$ and I_{max}, respectively, should be as large as possible. The phase shift between the armature current and the open-circuit EMF must be zero ($\psi = 0$). In general, the maximum amplitude of the armature current is limited by thermal constraints.

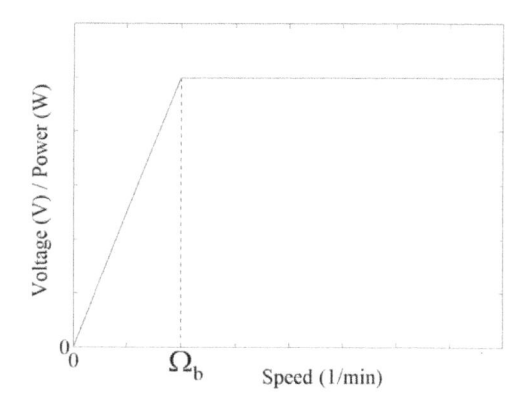

a) "Voltage/speed" or "power/speed" envelope

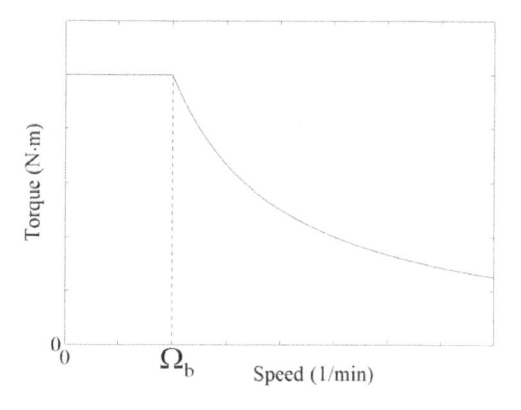

b) "Torque/speed" envelope

Figure 1.5. *"Voltage/speed", "power/speed" and "torque/speed" envelopes for electromechanical drives*

By adopting this strategy, the modulus of the total armature flux linkage is constant, as well as the torque. They are given by:

$$\Phi_{arm} = \sqrt{\Phi_{exc\,max}^2 + (L \cdot I_{max})^2} \qquad [1.6]$$

$$\Gamma_{em\,max} = p \cdot \Phi_{exc\,max} \cdot I_{max} \qquad [1.7]$$

It follows that the modulus of the armature voltage V varies linearly with the rotational speed:

$$V = \omega \cdot \Phi_{arm} = p \cdot \Omega \cdot \sqrt{\Phi_{exc\,max}^2 + (L \cdot I_{max})^2} \qquad [1.8]$$

With these equations, we gain a better understanding of why the voltage (power) and torque envelopes in Figure 1.4 have their shapes at low speeds ($\Omega \leq \Omega_b$).

However, from the base speed Ω_b, which corresponds to the speed at which the maximum voltage is reached, the strategy for imposing the maximum torque no longer works. In order to operate above this speed, the total armature flux linkage must be reduced. It is therefore necessary to implement flux-weakening strategies.

The following equation gives the expression of the total armature flux linkage in the general case:

$$\Phi_{arm} = \sqrt{(\Phi_{exc} - L \cdot I \cdot \sin(\psi))^2 + (L \cdot I \cdot \cos(\psi))^2} \qquad [1.9]$$

This shows that flux weakening can be achieved in several different ways. To reduce the flux, it is possible to adjust the modulus of the armature current I, its phase shift with respect to the EMF ψ or even the excitation flux Φ_{exc}. In the work by Amara et al. (2019), different strategies for weakening the flux have been studied. It was found that the best flux weakening strategy depends on the value of the normalized synchronous inductance L_n, which is defined by:

$$L_n = \frac{L \cdot I_{max}}{\Phi_{exc\,max}} \qquad [1.10]$$

If the value of the normalized synchronous inductance is greater than or equal to unity, then extending the operational range beyond the base speed does not require the excitation flux to be controlled. The contribution coming from hybrid excitation towards the weakening of the flux is in this case nil. If, on the contrary, the value of the normalized synchronous inductance is less than unity, the contribution coming from the hybrid excitation towards the extension of the operational range is undeniable. Figure 1.6 shows the power characteristics for two machines with the same normalized inductance value which is less than unity ($L_n = 0.5$). One of them uses permanent magnet excitation while the other uses hybrid excitation. This figure illustrates the contribution coming from hybrid excitation.

From a sizing point of view, it is preferable to have a normalized synchronous inductance value less than unity, whenever this is possible. Having $L_n < 1$ allows for better sizing of the converter-machine assembly. The apparent power of the converter to be sized is closer to the maximum active power of the machine in this case. However, in order to avoid any electromagnetic limitations when broadening the operational range, the hybrid excitation must reduce the excitation flux, so that the minimum value of k_f [1.3] is less than or equal to L_n.

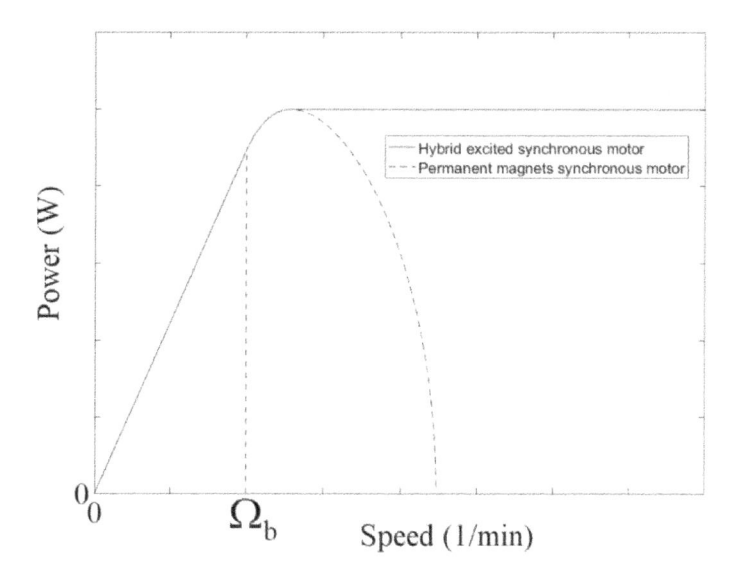

Figure 1.6. *Power characteristics for a machine with $L_n = 0.5$, with and without hybrid excitation*

This is one of the advantages of hybrid excitation, which allows the converter-machine assembly to be sized as accurately as possible, while still allowing the operational range to be extended beyond the base speed. We will return to the contribution from hybrid excitation for synchronous motors in more detail in Chapter 2.

1.2.2. Generation mode operation

There are three different cases in which an alternator may be used in the operation of a synchronous machine. These are (Figure 1.7):

– an alternator connected directly to the electrical grid (Figure 1.7(a));

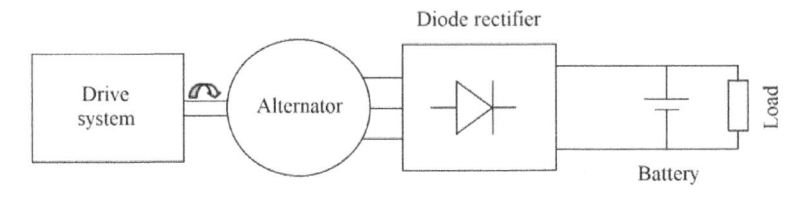

– an alternator connected to a diode rectifier (Figure 1.7(b));

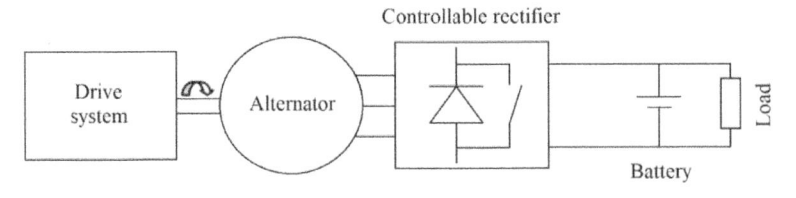

– an alternator connected to a controllable rectifier (Figure 1.7(c)).

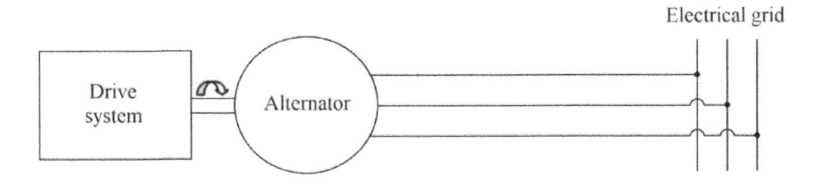

a) Alternator connected to the electrical grid (constant speed).

b) Alternator connected to a diode rectifier.

b) Alternator connected to a controllable rectifier.

Figure 1.7. *Example applications of alternators*

The first case corresponds to constant-speed operations, in which the electrical grid imposes the frequency. Of all the studies conducted on hybrid excited machines that use this type of application, two predominant motivations can be identified:

– improving the performance of conventional wound field excitation alternators (Fukami et al. 2010; Yamazaki et al. 2012);

– investigating the performance of hybrid excited alternators used in this type of application (Ammar et al. 2012; Kamiev et al. 2012, 2013, 2014; Kamiev 2013).

Although not fundamentally contradictory, these two motivations reflect different ambitions and therefore give rise to different approaches. These two motives can also be found in the other application cases.

For the first category, it is a question of starting from an existing wound field excitation alternator for which the addition of permanent magnets allows for the performance to be improved, increasing the amount of power delivered and also increasing the efficiency (Fukami et al. 2010; Yamazaki et al. 2012).

The second, and more general category, allows for a better understanding of the contribution that hybrid excitation brings when used in this type of application. They begin with qualitative (Ammar et al. 2012; Kamiev et al. 2012, 2014; Kamiev 2013) or quantitative (Kamiev et al. 2013), specifications, and then explore the performance of hybrid excited machines.

The second application case typically corresponds to alternators used in internal combustion motor vehicle applications (Bürger 1999), but not exhaustively. This type of association can also be found in the generation of onboard aircraft electrical power (Nasr 2017). As before, we can find studies where hybrid excitation is sought to improve the performance of existing wound excitation alternators (Albert 2004), and studies are now starting to integrate hybrid excitation, beginning with the design phase (Takorabet 2008; Nasr 2017).

Prior to the development of purely electric vehicles, the automotive industry experienced a period during which the onboard electrical power of internal combustion vehicles increased considerably, to accommodate the presence of more and more electrical equipment (Albert 2004). This trend is also present in the aviation industry (Roboam et al. 2012).

Hybrid excited alternators have been studied to address this issue (Albert 2004; Takorabet 2008). The interest in hybrid excitation for the "variable-speed alternator/diode rectifier" associations (Figure 1.7(b)) is trivial. Indeed, in this case, hybrid excitation allows for the DC bus voltage to be controlled, without requiring the use of a controllable rectifier (Amara et al. 2010) (Figure 1.7(c)), while the use of permanent magnets ensures high efficiency over a broad operational range.

The third application case corresponds to the reciprocal operation of the motoring mode operation, discussed previously in section 1.2.1. As for the motoring mode operation, the optimal control strategy depends on the value of the normalized inductance in the direct axis.

Overall, whether for motoring or generating operations, the use of hybrid excitation allows for greater flexibility in the design and the operation of synchronous machines. The permanent magnets allow them to operate with very high efficiency. Wound field excitation makes it possible to control the air-gap magnetic flux, thus making it possible to adapt these machines to very demanding specifications. However, this comes at the cost of a greater complexity when compared to single excitation synchronous machines, whether they use wound field excitation or permanent magnets.

1.3. Hybrid excited structures

The principle of hybrid excitation allows for a wide variety of structures to be realized, and so many criteria can then be used when classifying hybrid excited synchronous machines. The classical criteria used in the classification of other types of electrical machines are often used. For example, we have:

1) radial or axial flux machines (for rotational machines);

2) machines with a 2D (magnetic flux circulating mainly in one plane) or a 3D (magnetic flux circulating in three dimensions) structure;

3) rotary and linear machines.

Several publications have been devoted to classifying synchronous hybrid excited structures (Amara et al. 2001, 2004a, 2007; Hlioui et al. 2013; Kamiev 2013; Yang et al. 2016; Asfirane et al. 2019; Zhu and Cai 2019). Therefore, we will not go into detail on the great variety of hybrid excited structures, but instead propose an updated synthesis, highlighting the strong trends, while citing a large number of references. The reader can refer to them to consolidate their understanding and clarify the elements presented in this section.

1.3.1. *Classification criteria*

There are many criteria used in the classification but, given the principle that hybrid excitation uses two sources of magnetic excitation flux (Figure 1.1), i.e. permanent magnets and excitation coils, two criteria seem to be better suited:

1) By analogy with electric circuits, the way in which the two excitation flux sources are combined: series hybrid excited synchronous machines (SHESMs), and parallel hybrid excited synchronous machines (PHESMs) (Amara 2001).

2) The localization of excitation flux sources within the machine: both sources in the stator, both sources in the rotor, or a mixed localization. By mixed localization, we mean that one field source (wound field excitation coils or permanent magnets) is located within the rotor, or within the stator, and the other field source is in the stator or the rotor, respectively. Having wound field excitation coils in the stator is preferred to avoid sliding contacts (Amara et al. 2011; Kamiev 2013).

The first criterion makes it possible to situate hybrid excited machines in the general, broader, framework of magnetic circuits (Amara 2001), with an important influence on the efficiency of excitation flux control. It is a criterion that can be qualified as fundamental when compared with the second one, which is more in line with technological and industrial considerations.

Indeed, the location of the sources can have a strong impact on the manufacturing complexity of the machines. Having all the excitation sources, as well as the armature windings, in the stator, which allows for a completely passive moving part, simplifies the construction and offers functional advantages. Cooling, for example, is simplified. Flux-switching hybrid excited synchronous machines correspond to this description, and this is not unrelated to the interest shown in research laboratories (Yang et al. 2016; Asfirane et al. 2019; Zhu and Cai 2019) and (Hoang et al. 2007; Sulaiman et al. 2011; Wang and Deng 2012a, 2012b).

We now will present an updated synthesis of hybrid excited structures by classifying them according to the first criterion, although the others will not be neglected.

1.3.1.1. *Series hybrid excited synchronous machines (SHESMs)*

Figure 1.8 illustrates the principle of series hybrid excitation. Figure 1.8(c) shows a series hybrid excited synchronous machine (Mizuno 1997; Fodorean et al. 2007).

Figure 1.8(a) shows a schematic diagram of series hybrid excitation, and Figure 1.8(b) shows the equivalent magnetic circuit. The two excitation sources are in series. The relative permeability of the ferromagnetic parts is assumed to be infinite. P_{AG} and P_{PM} represent the permeances of the air-gap (AG) and the permanent magnet (PM), respectively.

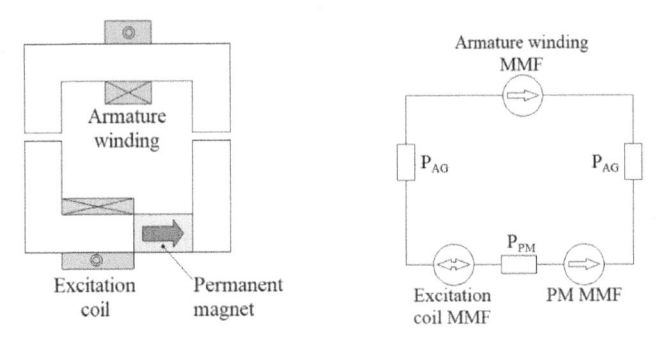

a) Schematic diagram.   b) Equivalent magnetic circuit.

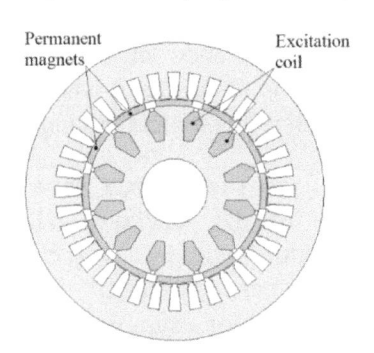

c) Example of a series hybrid excited synchronous machine (Mizuno 1997; Fodorean et al. 2007).

Figure 1.8. *Series hybrid excitation*

Figure 1.9 illustrates how the strengthening, or weakening, of the excitation field of the magnet is achieved by supplying the excitation coil with a current in either direction.

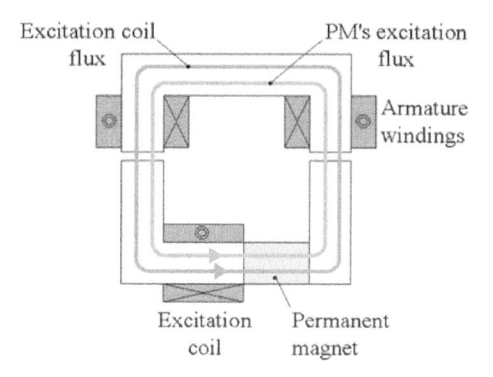

a) Strengthening the excitation flux of the magnet.

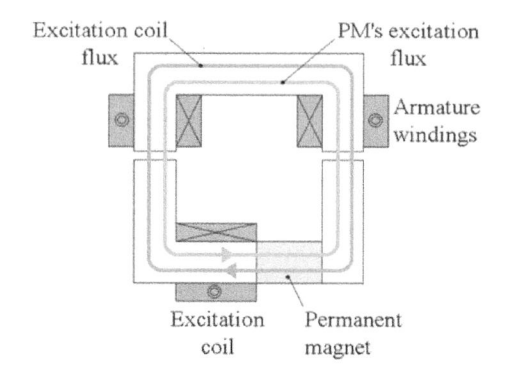

b) Weakening the excitation flux of the magnet

Figure 1.9. *Control of the total excitation flux for SHESMs*

In SHESMs, since the excitation flux created by the excitation coils must pass through the PMs (Figure 1.9) which have low relative permeability ($\mu_{rPM} \approx 1$), i.e. a low permeance P_{PM} (Figure 1.8(b)), the flux control capability is expected to be less effective than that found in parallel hybrid excited structures. In addition, the risk of demagnetizing the PMs raises concerns. However, since the excitation fluxes created by either source follow the same path, the iron losses decrease when the excitation flux of the PMs is weakened, which is not the case for all hybrid excited machines.

The structure shown in Figure 1.8(c) is a 2D structure, in which both excitation sources are found in the rotor. The structure shown in Figure 1.2 is also a SHESM, although in 3D and with the two sources located in the stator. Other SHESM structures will be illustrated in section 1.3.2.

1.3.1.2. *Parallel hybrid excited synchronous machines (PHESMs)*

Figure 1.10 illustrates the principle of parallel hybrid excitation. Figure 1.10(c) shows a parallel hybrid excited synchronous machine (Akemakou and Phounsombat 2000).

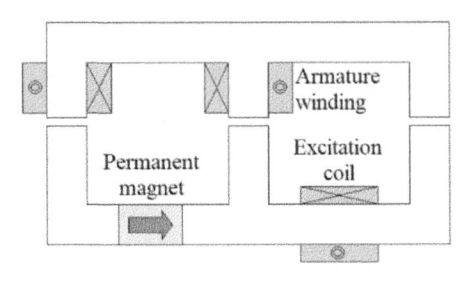

a) Schematic diagram.

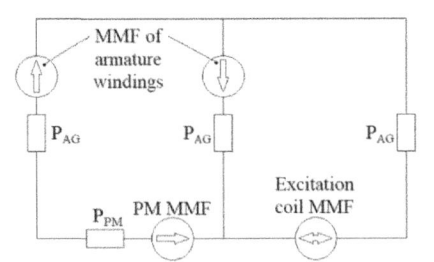

b) Equivalent magnetic circuit.

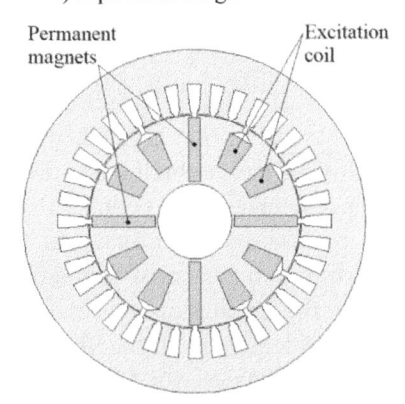

c) Example of a parallel hybrid excited synchronous machine (Akemakou and Phounsombat 2000).

Figure 1.10. *Parallel hybrid excitation*

Figure 1.10(a) shows a schematic diagram of parallel hybrid excitation, and Figure 1.10(b) shows the equivalent magnetic circuit. Just as for SHESMs, the relative permeability of the ferromagnetic parts is assumed to be infinite. The two excitation sources are in parallel. The main wound excitation coil flux does not pass through the permanent magnet.

Figures 1.11(a) and 1.11(b), respectively, illustrate how the strengthening and the weakening of the excitation field magnet are achieved via the excitation coil supply. These figures show the main paths of the excitation fluxes produced by the two sources.

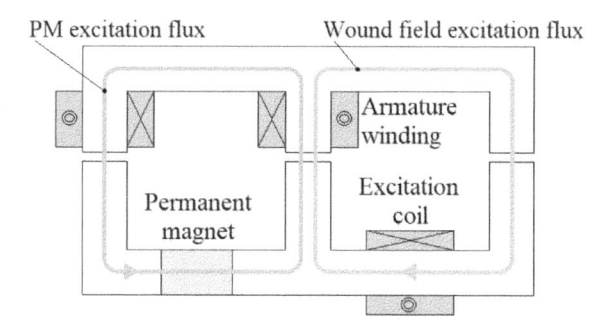

a) Strengthening the excitation flux of the magnet.

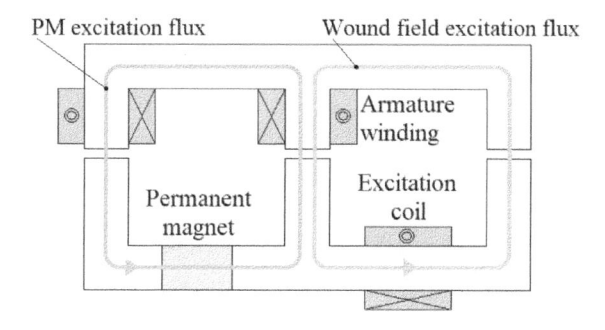

b) Weakening the excitation flux of the magnet.

Figure 1.11. *Control of the total excitation flux for PHESMs*

Figure 1.12 shows the auxiliary paths, which are less efficient for inducing a voltage in the armature windings. This is the main drawback of PHESMs, but the control of the total excitation flux is more efficient overall than for SHESMs, however (Amara et al. 2011; Hua and Zhu 2020).

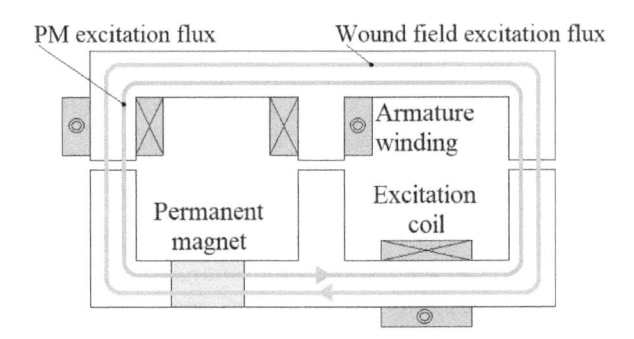

Figure 1.12. *Auxiliary flux paths created by the excitation sources in PHESMs*

As with the SHESM shown in Figure 1.8(c), the PHESM shown in Figure 1.10(c) has excitation sources located at the rotor. We will see in section 1.3.2 that other configurations are possible for both SHESMs and PHESMs.

1.3.1.3. *SHESMs versus PHESMs*

It is reasonable to ask about the effectiveness when it comes to controlling the total excitation flux in the two classes defined above since this is one of the main reasons for developing hybrid excited synchronous machines.

Interesting studies comparing this effectiveness can be found in works by Amara et al. (2011) and by Hua and Zhu (2020). All these studies show the superiority of PHESMs over SHESMs. This is mainly due to the low permeability of permanent magnets, which naturally limits the efficiency of the excitation flux control for SHESMs. These studies also highlight the risk of demagnetization of permanent magnets, which is greater for SHESMs. However, SHESMs appear to provide a greater power density (Hua and Zhu 2020) since, for the same volume of permanent magnets, PHESMs suffer from a lower utilization of the magnets, related to the auxiliary fluxes that reduce their efficiency (Figure 1.12).

Given the advantages of PHESMs over SHEMSs, especially regarding the efficiency when controlling the total excitation flux, most of the structures studied in the scientific and technical literature are PHESMs. In particular, flux-switching PHESMs are currently more favored (Hoang et al. 2007; Sulaiman et al. 2011; Wang and Deng 2012a, 2012b; Yang et al. 2016; Asfirane et al. 2019; Zhu and Cai 2019).

1.3.2. *Structures and classification*

The aim here is not to carry out an exhaustive review of HESMs, but rather to show the diversity of the structures studied (Amara et al. 2001, 2004a, 2007; Hlioui et al. 2013; Kamiev 2013; Yang et al. 2016; Asfirane et al. 2019; Zhu and Cai 2019).

Figure 1.13 shows a SHESM with a 2D structure, where all of the excitation sources, including the armature windings, are located within the stator (Leonardi et al. 1996). The rotor is completely passive here.

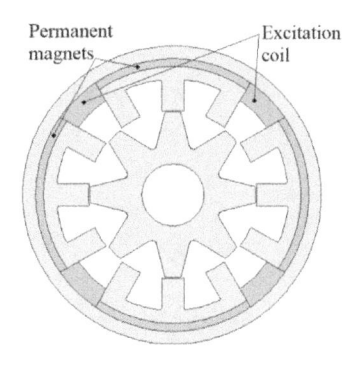

Figure 1.13. *2D structure of a SHESM with a passive rotor (Leonardi et al. 1996)*

Figure 1.14 shows two perspectives of the rotor found in a PHESM. This is the claw-pole alternator used in automobiles, with magnets inserted in between the claws (Henneberger et al. 1996; Bürger 1999; Richard and Dubel 2007; Boldea 2016). This configuration allows for a greater amount of power to be converted by the machine. This machine is used as a starter-alternator in motor vehicles with an internal combustion engine (Richard and Dubel 2007; Boldea 2016), which are called micro-hybrids. To our knowledge, this is the only industrial development of an HESM to be used by the general public. The alternator mode of this system could correspond to the use cases featured in Figure 1.7(b) and 1.7(c).

This is a PHESM with a 3D structure, with the excitation sources localized within the rotor. Figure 1.15 shows two PHESMs with 3D structures, based on the same hybrid excited configuration. The magnetic flux flows in all three dimensions, but with a preferential direction at the air-gap. Both structures can be classified as radial flux (Figure 1.15(a)) (Tapia et al. 2003) and axial flux

(Figure 1.15(b)) (Aydin et al. 2002; Aydin 2004) machines, respectively. The annular excitation coil, which is not shown in Figure 1.15(b), is located within the stator between the two rotor disks.

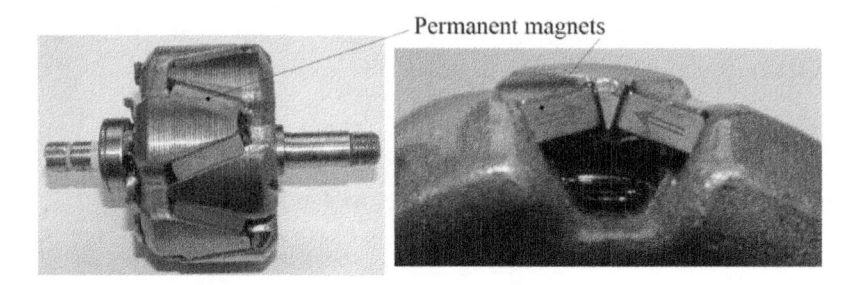

Figure 1.14. *Claw-pole rotor used in hybrid excitation (Albert 2004)*

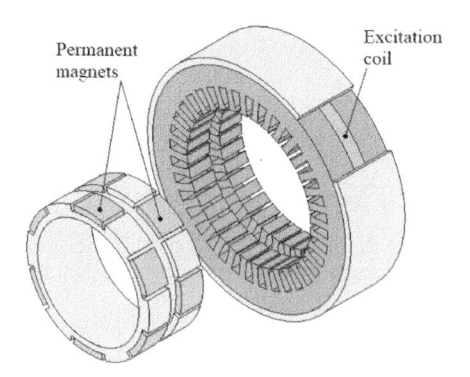

a) PHESM with a radial flux in the air-gap (Tapia et al. 2003).

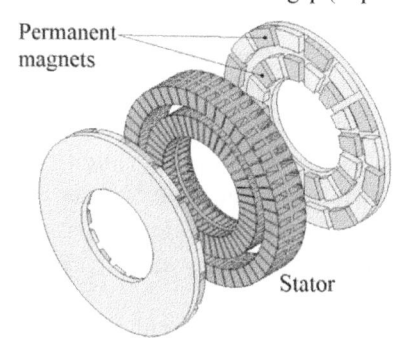

b) PHESM with an axial flux in the air-gap (Aydin et al. 2002; Aydin 2004).

Figure 1.15. *PHESM with excitation coils located within the stator*

For these two machines, the excitation coils are located within the stator and the permanent magnets in the rotor. The presence of the excitation coils in the stator avoids the use of sliding contacts.

Figure 1.16 shows another PHESM, where the permanent magnets are in the rotor and the annular field coils in the stator (Amara et al. 2009). This is also a 3D structure, with a radial flux in the air-gap of the active part. This figure also shows the main flux paths of the excitation fluxes coming from both sources, i.e. the permanent magnets and the excitation coils. The main flux of the permanent magnets is bipolar, with the creation of a north and a south pole at the air-gap of the active part. Each of the two annular excitation coils act solely on one type of magnetic pole.

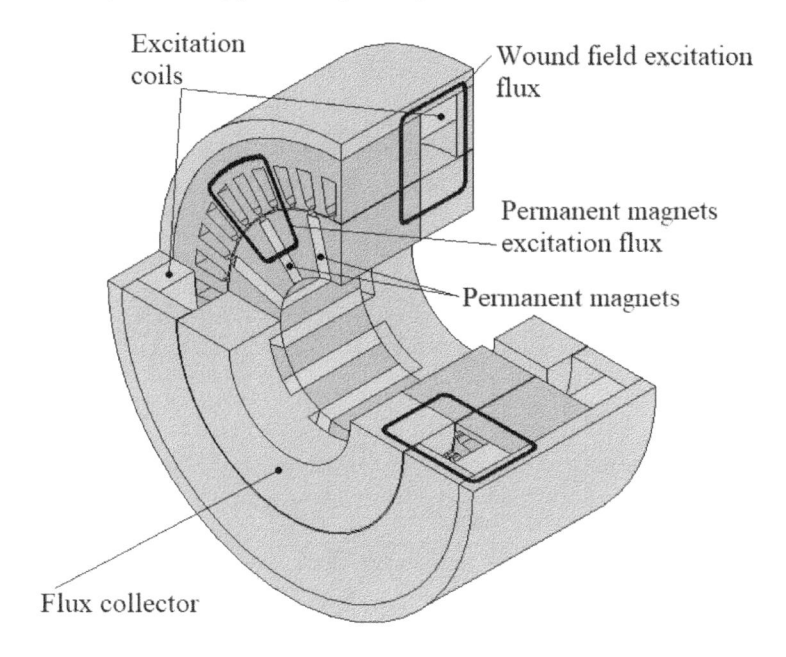

Figure 1.16. *3D cut view of a bipolar PHESM with excitation coils located in the stator*

Figure 1.17 shows the auxiliary paths of the excitation flux created by the permanent magnets. The contribution coming from these magnetic fluxes in the conversion is less efficient than that from the main bipolar path (Figure 1.16). The path illustrated in Figure 1.17(c) corresponds to a leakage flux, which does not contribute anything towards the conversion. The presence of these paths reduces the efficiency when using permanent magnets, as discussed in section 1.3.1.3.

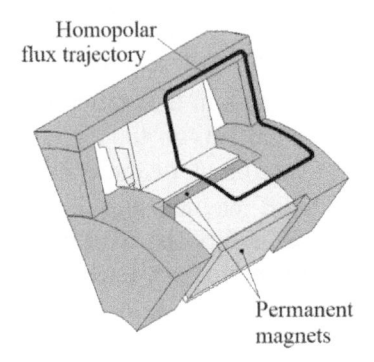

a) First homopolar path created by the permanent magnets.

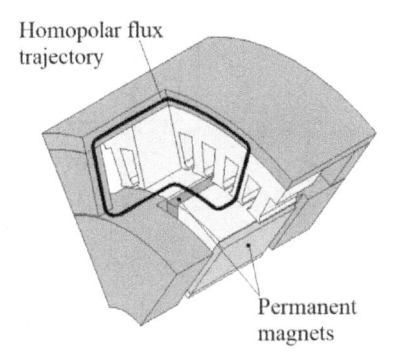

b) Second homopolar path created by the permanent magnets.

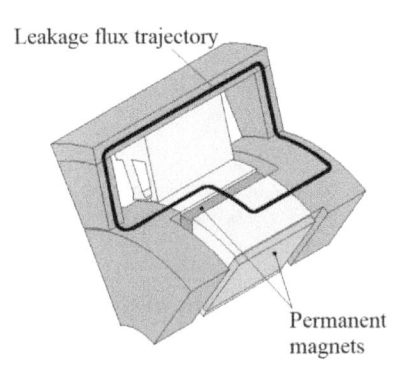

c) Path of the leakage flux from the permanent magnets.

Figure 1.17. *Auxiliary paths of the excitation flux from the permanent magnets*

The structure in Figure 1.16 is called bipolar because the wound excitation flux acts on both the north and south magnetic poles under the active part of the machine (Vido et al. 2005a, 2005b). The parts called "flux collectors" (Figure 1.16), located at the axial ends of the rotor, are offset by one pole pitch for this PHESM.

Figure 1.18 shows a homopolar version, where the flux collectors in the rotor are aligned. The wound excitation flux acts on only one of the two poles.

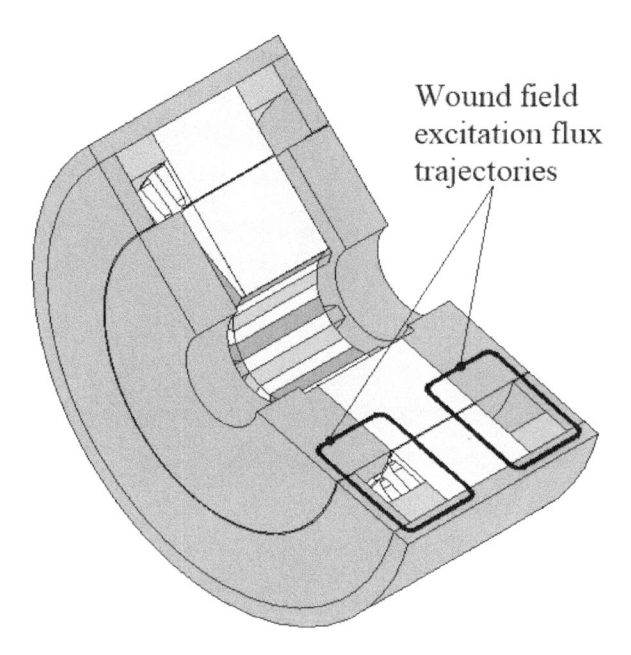

Figure 1.18. *3D cut view of a homopolar PHESM with excitation coils located in the stator*

For HESMs with a 3D structure, the magnetic fluxes coming from the different sources flow in three dimensions, and it is necessary to find a compromise between several aspects when it comes to choosing the materials for these machines, in particular their shape, i.e. solid or laminated. It is necessary to provide the path of least reluctance to the useful excitation flux, reduce the less useful auxiliary paths, take into account iron losses induced in these materials, and consider the mechanical design constraints.

For 2D structures, laminated materials are more appropriate to reduce iron losses, but again here mechanical constraints must be considered.

Figure 1.19 shows a flux-switching PHESM with a 2D structure (Nasr 2017). The rotor is completely passive. Flux-switching HESMs have a growing interest because of their effective functional characteristics, along with having all of the magnetic flux sources gathered within a single armature, usually the stator. However, the location of all the flux sources within a single armature gives rise to disadvantages such as the reduced space available for the ferromagnetic circuit, the amplification of magnetic saturation effects, and limits the generation of the torque, or the forces in the case of linear machines (Amara et al. 2013). In addition, an induced voltage is encountered in the excitation coil of these machines, which can disrupt its supply (Sun and Zhu 2020).

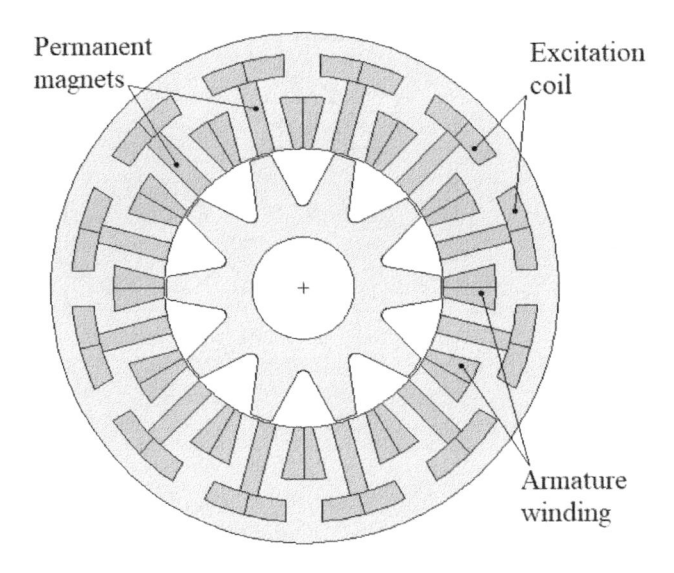

Figure 1.19. *Flux-switching PHESM with a 2D structure (Nasr 2017)*

To address the problem of limited space available for the ferromagnetic circuit, researchers have proposed to distribute the sources over two independent armatures (Hua and Zhu 2017). Figure 1.20 shows a flux-switching SHESM with two static and one mobile armature. The mobile armature is included between the two static armatures, which turns out to be a drawback in the design of this structure.

Solutions based on 3D structures have also been proposed. Figures 1.21 and 1.22 show structures where the relocation of the wound excitation circuit partially remedies this limitation. For the structure in Figure 1.21, the wound excitation circuit is relocated radially towards the outer periphery of the machine (Dupas et al. 2016).

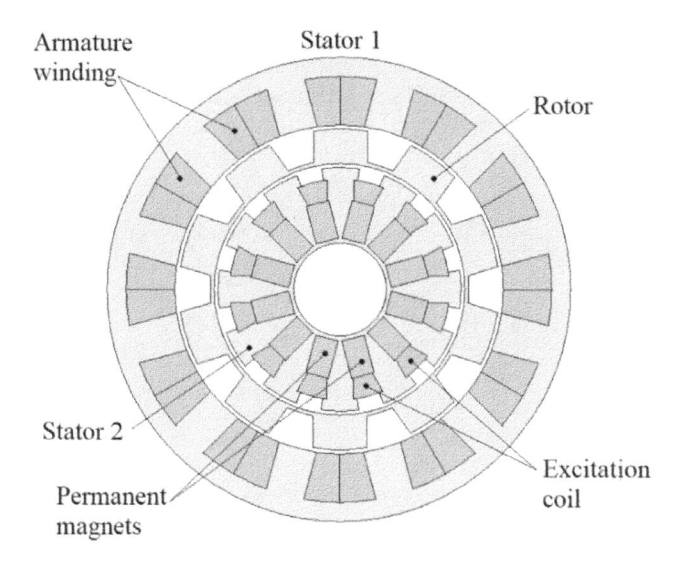

Figure 1.20. *2D structure of a flux-switching SHESM with three armatures (two stator armatures and one mobile armature) (Yang et al. 2016)*

For the structure in Figure 1.22, the wound excitation circuit is divided into two parts located at the axial ends of the machine (Li et al. 2021). Both structures (Figures 1.21 and 1.22) are based on the same hybrid excitation principle.

It should be noted that these structures also avoid the problem of an induced voltage in the excitation coils (Sun and Zhu 2020; Hlioui et al. 2022). This voltage becomes all the more important when the speed is high. This is a problem that is not only unique to flux-switching HESMs but also concerns hybrid excited machines with distributed excitation coils, such as the structure shown in Figure 1.8(c).

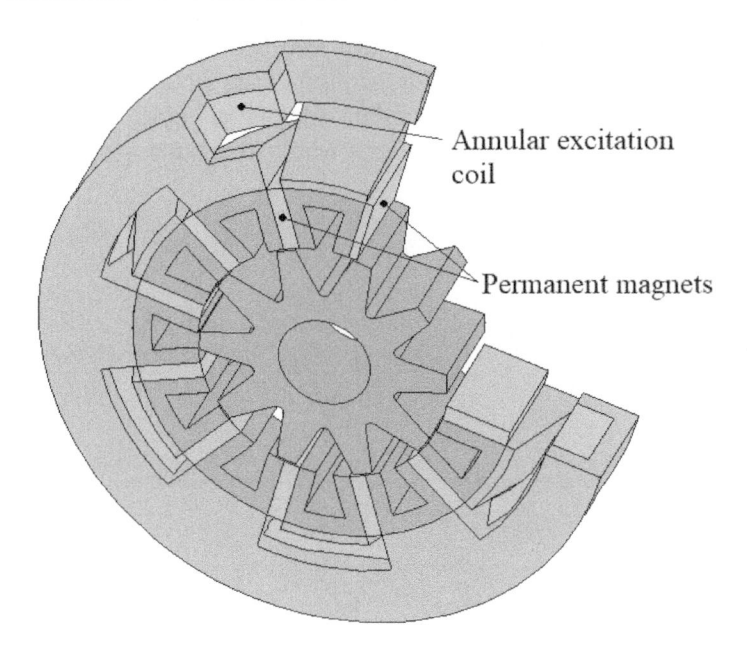

Figure 1.21. *3D cut view of a flux-switching PHESM structure (Dupas et al. 2016)*

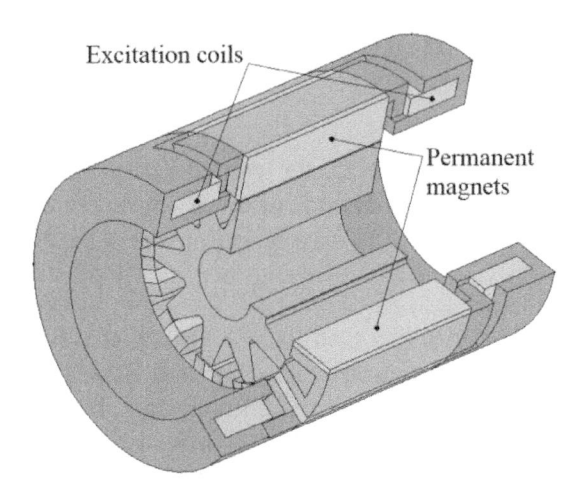

Figure 1.22. *3D cut view of a flux-switching PHESM structure (Li et al. 2021)*

The machines in Figures 1.21 and 1.22 have a structure that naturally avoids this problem, with completely annular excitation coils (Ben Ahmed 2006). For HESMs with distributed excitation coils, alternative solutions have been proposed to reduce the induced voltage (Wu et al. 2019a; Hlioui et al. 2022).

The induced voltage is essentially due to the variation in the reluctance experienced by the excitation coils. This is a similar problem to that of the cogging torque (or force) found in permanent magnet machines. It is therefore natural that researchers have investigated methods used to reduce the cogging torque (or force), such as skewing (Wu et al. 2019a).

In the work by Hlioui et al. (2022), an alternative solution based on the use of damper windings allows for the magnetic field responsible for this induced voltage to be reduced. This is a short-circuited winding that is distributed in the same way as the excitation coils.

All the structures listed above are rotational machines. The principle of hybrid excitation is also applicable to linear machines (Hlioui et al. 2022). Figure 1.23 shows an example (Zeng and Lu 2018). This is a linear flux-switching hybrid excited machine, for which the magnetic flux sources are distributed over two armatures, with a third passive armature.

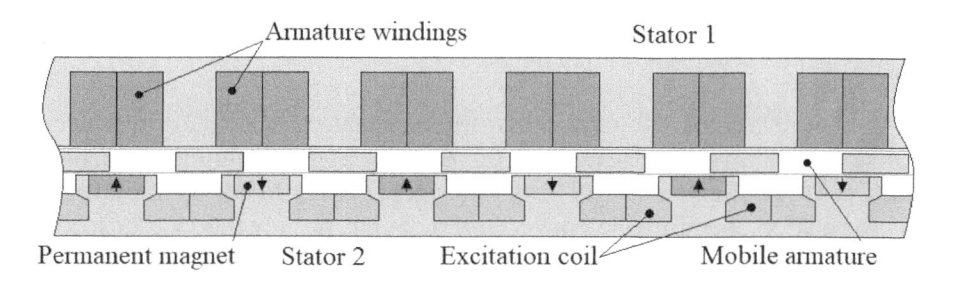

Figure 1.23. *Linear flux-switching SHESM (Zeng and Lu 2018)*

This section illustrates the great diversity of HESMs, which is concomitant with the diversity of single excitation (or mono-excitation) synchronous machines and is further enhanced by the flexibility of the two excitation sources.

1.4. Conclusions and perspectives

The possibilities offered by the principle of hybrid excited, both from the point of view of combining the advantages of permanent magnet machines (effective efficiency) and wound field excitation machines (flexibility of operation), in addition to that of realizable structures, make HESMs an interesting research topic that is relevant for various applications (Yang et al. 2016; Wardach et al. 2020; Hlioui et al. 2022).

It is difficult to exhaustively list each and every study dedicated to HESMs, and this is not the purpose of this chapter, but the numerous references associated with this chapter will allow readers to verify, clarify and perfect their understanding of the concepts presented.

In general, studies on HESMs are inseparable from research on synchronous machines. It will always be relevant to ask whether adding some quantity of permanent magnets to a wound field excitation machine, or conversely by adding a wound field excitation circuit to a permanent magnet machine, would improve the performance for a reasonable cost.

This fact is all the more topical as society continues to search for clean and sustainable means of energy conversion, and electrical machines are an important factor in these energy conversion chains. This is promising for research on actuators and electrical machines in general and, in particular, on hybrid excited synchronous machines.

We can therefore reference the work by Yang et al. (2012), in which a synergy between the advantages of HESMs and synchronous memory machines (Christophe et al. 2014) was sought. For the motor operation, only the association "controllable converter/motor" has been mentioned in section 1.2.1. We could very well consider, as the reciprocating application of the grid-connected alternator case (Figure 1.7(a)), the grid-connected motor operation, where a motor directly connected to the grid with line-starting capability would be required (Amara et al. 2021b; Hlioui et al. 2022). Figure 1.24 shows a HESM with such a capability.

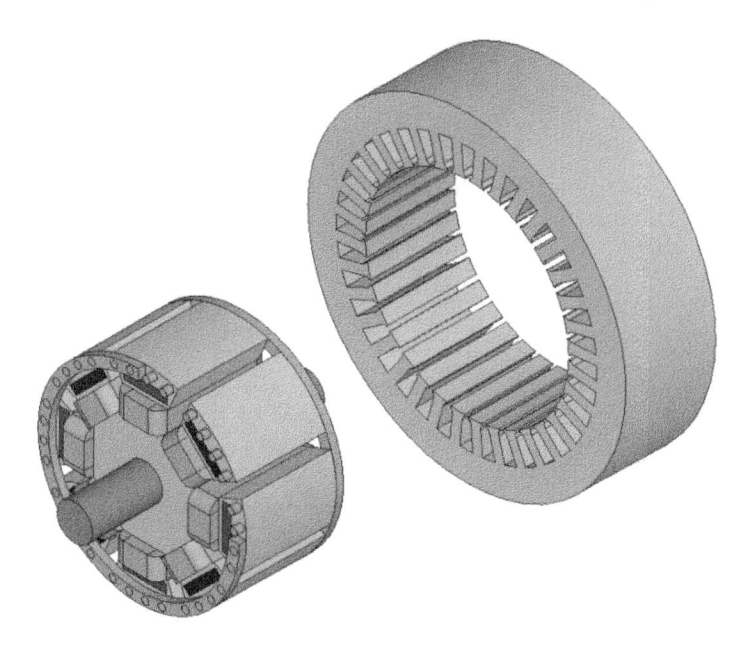

Figure 1.24. *SHESM with a squirrel-cage rotor (Amara et al. 2021b; Hlioui et al. 2022)*

2

Control of Hybrid Excited Synchronous Machines

2.1. Introduction

This chapter is devoted to studying the control laws of three-phase synchronous machines that operate in motor mode. These laws are studied in order to help a designer of an electric drive make decisions for any application use.

The term "original study" mentioned in the Introduction emphasizes that this presentation is from the point of view of a designer of electrical machines, rather than that of control laws. Most books devoted to the control of electric machines (Yamamura 1986; Vas 1992; Lacroux 1994; Mohan et al. 1995; Grellet and Clerc 1997; Leonhard 1997; Bose 2001; Louis 2010, 2011; De Doncker et al. 2011) are written by specialist researchers in the field, or in power electronics. Emphasis is placed on implementation aspects, in the context of power electronics converters/electric machine associations. The dynamic aspects often take an important place. This is not the case in this chapter, which is devoted to steady-state operations (it will be assumed that the power converter allows us to impose the desired control quantities).

The mathematical developments here will allow for a concise understanding of the laws and their foundations and will be detailed with a focus on the formal analytical approach. To not make this chapter too heavy

---

For a color version of all the figures in this chapter, see www.iste.co.uk/amara/hybrid.zip.

going, the lengthy mathematical developments will be available in a downloadable appendix (https://1drv.ms/b/s!AogSAGtYvycUkHIbc3FbsPoSQjJt?e=U9p85C). However, we hope that the reader will be able to reproduce these developments, should it be necessary. The laws will be studied according to the machine characteristics, i.e. for different sets of parameters of the circuit models that will be used for this purpose. In order to illustrate the contribution coming from hybrid excitation, the laws studied are for machines with a fixed excitation flux (permanent magnet machines) and variable-excitation flux (machines with wound or hybrid excitation). Synchronous machines with variable reluctance will also be studied, even if this is a somewhat marginal topic.

2.2. Modeling of hybrid excited synchronous machines

The study of the control laws is based on the classical three-phase synchronous machine models expressed in the "Park" reference frame (Morimoto et al. 1994; Fernandez-Bernal et al. 2001), in a steady-state and with a balanced three-phase supply. The currents entering the machine are counted positively (the motor convention), and the magnetic circuit is assumed to be linear. Moreover, we place ourselves within the first harmonic hypothesis framework, in which we assume that each electromagnetic quantity behaves sinusoidally. Therefore, each quantity expressed in the "Park" formalism is constant.

Figure 2.1 presents equivalent electrical circuits of hybrid excited synchronous machines. Compared to wound excitation synchronous machines, the total EMF (electromotive force) which, in the stator circuit, is along the transverse axis (q-axis) is the sum of the EMF contributions from the permanent magnets and the wound inductor. This model integrates the iron and Joule losses into it, which occur in synchronous machines.

Figure 2.2 shows the vector diagram of the electromagnetic quantities (flux, EMF, voltage, current) in the "Park" reference frame. By convention, the angles ψ and δ, as they appear in this figure, are counted negatively and positively, respectively. The vectors are assumed to rotate in the direct (counterclockwise) direction. Table 2.1 lists the symbols used in Figures 2.1 and 2.2.

On the one hand, the Joule losses in the armature and excitation circuits are represented by the resistances R_s and R_e, respectively. On the other hand, the core losses, which are located in the stator, are represented by the resistance R_f. The latter is caused by flux variations in the stator, and the resistance R_f is therefore placed in parallel with the voltage sources induced by the fluxes in the circuits along the direct (d-axis, Figure 2.1(a)) and transverse (q-axis, Figure 2.1(b)) axes.

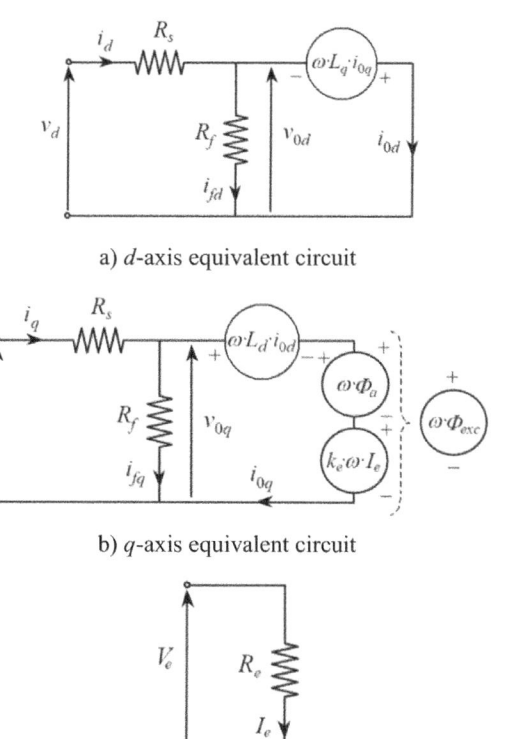

a) d-axis equivalent circuit

b) q-axis equivalent circuit

c) Wound field excitation equivalent circuit

Figure 2.1. Equivalent electrical circuits for hybrid excited synchronous machines

This model is general in the sense that any model for the different types of synchronous machines, i.e. wound excited synchronous machines, permanent magnet synchronous machines and variable-reluctance synchronous machines, can be derived from it. Table 2.2 summarizes the necessary adaptations for each type of machine and presents the necessary values of the parameters and quantities that should be adopted when considering each type of machine.

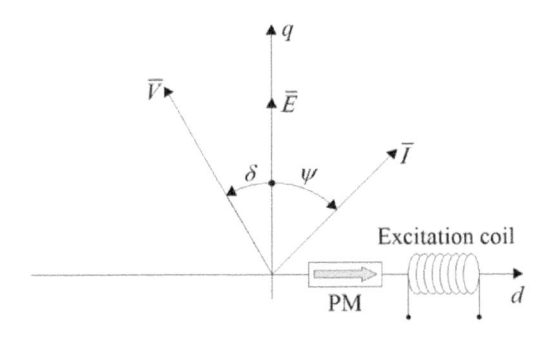

Figure 2.2. *Vector diagram of the main electromagnetic quantities of an HESM*

Symbols	Definitions and units
i_d, i_q	d- and q-axis components of the armature current (A)
I_e	Excitation current (A)
i_{fd}, i_{fq}	d- and q-axis components of the iron loss current (A)
v_d, v_q	d- and q-axis components of the terminal voltage (V)
V_e	Terminal voltage of wound field excitation circuit (V)
R_s	Armature winding resistance per phase (Ω)
R_f	Iron loss resistance (Ω)
R_e	Wound inductor resistance (Ω)
Φ_a	Permanent magnet excitation flux (Wb)
Φ_{exc}	Total excitation flux (Wb)
k_e	"Armature/wound inductor" mutual inductance (H)
L_d, L_q	d- and q- axis components of the armature inductance (H)
ω	Electrical pulse (rad/s)
ψ	"EMF/armature current" phase shift (rad)
δ	"EMF/armature voltage" phase shift (internal angle) (rad)

Table 2.1. *Quantities and parameters for the equivalent electrical circuits and the vector diagram*

Type of synchronous machine	Values of the quantities and parameters
Wound excitation machines	$\Phi_a = 0$ Wb
Permanent magnet machines	$k_e = 0$ H and no wound inductor circuit
Variable-reluctance synchronous machines	$\Phi_a = 0$ Wb, $k_e = 0$ H, and no wound inductor circuit

Table 2.2. *Values of the quantities and parameters for different types of synchronous machines*

2.2.1. *The nature of the equations*

The equations come from determining the armature current and voltage amplitudes, as well as the expressions for the power, losses and torque of a synchronous machine. These amplitudes allow us to verify that the established control laws respect the constraints that are imposed for the desired application. These equations are established in the "Park" reference frame using the equivalent electrical circuit models (Figure 2.1). It should be noted that the "Park" transformation, which we will adopt, is one in which the power is conserved (Chatelain 1989).

The equations are first established by considering the Joule and iron losses before neglecting them. Mechanical losses are not considered in our study, but they may be integrated into it as the methodology we will adopt allows for this, which we will see later in this chapter.

The typical application chosen when studying different control laws is that of electrical traction/propulsion (Figure 2.3). Figure 2.4 presents the characteristics required in such an application, which are quite general and shared by a wide range of applications (Multon et al. 1995).

Figure 2.3. *An electric vehicle*

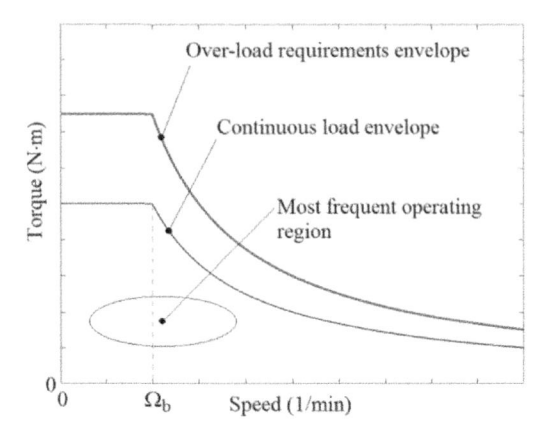

Figure 2.4. *"Torque/speed" characteristics for electric vehicle applications*

This application use is chosen because it requires a variable speed operation with a partial load. It is almost impossible to optimize the losses for each operating point, and therefore it is essential to include them in the control law study. The electric vehicle application is the most demanding in terms of control laws. For constant speed applications that have very little torque variation, designers manage to minimize losses by appropriate dimensioning.

Since full-load operations are not the most stressed operation within the operating cycle (Figure 2.4) of an electric vehicle, we will not study its efficiency. This operation is essentially dimensioning for the power supply of the electric machine, which justifies why we neglect the losses when studying the envelopes, and then take them into account for partial-load operations.

Elements provided in this chapter allow us to put the various studies conducted on the control laws of synchronous machines into perspective (Lajoie-Mazenc and Viarouge 1991; Multon et al. 1995; Soulard 1998; Soulard and Multon 1999; Amara 2001; De Doncker et al. 2011; Nam 2019).

2.2.1.1. *Model with losses*

If the Joule and iron losses are considered, the equations of the armature and inductor electrical circuits can be written as:

$$\begin{bmatrix} v_d \\ v_q \end{bmatrix} = R_s \cdot \begin{bmatrix} i_d \\ i_q \end{bmatrix} + \begin{bmatrix} v_{0d} \\ v_{0q} \end{bmatrix}$$

$$\begin{bmatrix} v_{0d} \\ v_{0q} \end{bmatrix} = \begin{bmatrix} 0 & -\omega \cdot L_q \\ \omega \cdot L_d & 0 \end{bmatrix} \cdot \begin{bmatrix} i_{0d} \\ i_{0q} \end{bmatrix} + \omega \cdot \Phi_{exc} \cdot \begin{bmatrix} 0 \\ 1 \end{bmatrix} \quad [2.1]$$

$$V_e = R_e \cdot I_e$$

The relationships between the different d- and q-axis components of the currents are given by:

$$\begin{bmatrix} i_d \\ i_q \end{bmatrix} = \begin{bmatrix} i_{fd} \\ i_{fq} \end{bmatrix} + \begin{bmatrix} i_{0d} \\ i_{0q} \end{bmatrix}$$

$$\begin{bmatrix} i_{fd} \\ i_{fq} \end{bmatrix} = \frac{1}{R_f} \cdot \left(\begin{bmatrix} 0 & -\omega \cdot L_q \\ \omega \cdot L_d & 0 \end{bmatrix} \cdot \begin{bmatrix} i_{0d} \\ i_{0q} \end{bmatrix} + \omega \cdot \Phi_{exc} \cdot \begin{bmatrix} 0 \\ 1 \end{bmatrix} \right) \quad [2.2]$$

Furthermore, from the vector diagram (Figure 2.2), the d- and q-axis components of the armature current and voltage can also be expressed in terms of the amplitudes I and V, and the phase shift angles with the EMF, ψ and δ (Table 2.1), by equations [2.3].

$$\begin{cases} i_d = -I \cdot \sin(\psi) \\ i_q = I \cdot \cos(\psi) \end{cases} \text{ and } \begin{cases} v_d = -V \cdot \sin(\delta) \\ v_q = V \cdot \cos(\delta) \end{cases} \quad [2.3]$$

We deduce that the expressions of the amplitudes of the armature current and voltage, as functions of the current components, are:

$$I = \sqrt{i_d^2 + i_q^2} = \sqrt{\left(i_{0d} - \frac{\omega \cdot L_q}{R_f} \cdot i_{0q}\right)^2 + \left(i_{0q} + \frac{\omega \cdot L_d \cdot i_{0d} + \omega \cdot \Phi_{exc}}{R_f}\right)^2}$$

$$V = \sqrt{v_d^2 + v_q^2} = \sqrt{\left(R_s \cdot i_d - \omega \cdot L_q \cdot i_{0q}\right)^2 + \left(R_s \cdot i_q + \omega \cdot L_d \cdot i_{0d} + \omega \cdot \Phi_{exc}\right)^2} \quad [2.4]$$

Since the "Park" transformation we have adopted preserves the power, the total power absorbed by the synchronous machine can be expressed as:

$$P = v_d \cdot i_d + v_q \cdot i_q + V_e \cdot I_e = \begin{bmatrix} R_s \cdot \left(i_d^2 + i_q^2\right) + R_e \cdot I_e^2 \\ + R_f \cdot \left(i_{fd}^2 + i_{fq}^2\right) \\ + \omega \cdot \left(\Phi_{exc} \cdot i_{0q} + (L_d - L_q) \cdot i_{0d} \cdot i_{0q}\right) \end{bmatrix} \quad [2.5]$$

Starting from the top expression in square brackets, the first term corresponds to the overall Joule losses, the second to the iron losses in the stator and the third to the mechanical power available on the machine shaft. The electromagnetic torque can then be expressed as:

$$\Gamma_{em} = \frac{P}{\Omega} = p \cdot \left(\Phi_{exc} \cdot i_{0q} + (L_d - L_q) \cdot i_{0d} \cdot i_{0q}\right) \quad [2.6]$$

where p is the number of pole pairs. Recall that, for synchronous machines: $\omega = p \cdot \Omega$, where Ω is the mechanical speed.

This model with losses is used when studying the operation on the entire "torque/speed" plane.

2.2.1.2. Lossless model

The lossless model is derived directly from the model with losses, by imposing: $R_s = 0\ \Omega$ and $R_f = +\infty$. We then have:

$$\begin{cases} i_{0d} = i_d \\ i_{0q} = i_q \end{cases} \text{ and } \begin{cases} v_{0d} = v_d \\ v_{0q} = v_q \end{cases}$$

We may rewrite the expressions for the armature current and voltage amplitudes [2.4] as a function of each other, as follows:

$$I = \sqrt{\left(\frac{V}{L_d \cdot \omega}\right)^2 \cdot \left(\left(\frac{\sin(\delta)}{\rho}\right)^2 + \cos^2(\delta)\right) + \left(\frac{\Phi_{exc}}{L_d}\right)^2 - \frac{2 \cdot \Phi_{exc} \cdot V \cdot \cos(\delta)}{L_d^2 \cdot \omega}} \quad [2.7]$$

$$V = \omega \cdot \sqrt{\left(L_q \cdot I \cdot \cos(\psi)\right)^2 + \left(\Phi_{exc} - L_d \cdot I \cdot \sin(\psi)\right)^2}$$

and also, to express the angles ψ and δ as functions of each other:

$$\operatorname{tg}\delta = \frac{L_q \cdot I \cdot \cos(\psi)}{\Phi_{exc} - L_d \cdot I \cdot \sin(\psi)}$$

$$\operatorname{tg}\psi = \rho \cdot \left(\frac{\Phi_{exc} \cdot p \cdot \Omega - V \cdot \cos(\delta)}{V \cdot \sin(\delta)} \right)$$

[2.8]

The parameter ρ appearing in equations [2.7] and [2.8] corresponds to the saliency ratio, which is defined as follows:

$$\rho = \frac{L_q}{L_d} \qquad [2.9]$$

Finally, the expression for the torque can be rewritten as a function of either the armature current or the armature voltage magnitude:

$$\Gamma_{em} = \frac{P}{\Omega} = p \cdot I \cdot \cos(\psi) \cdot \left(\Phi_{exc} - (L_d - L_q) \cdot I \cdot \sin(\psi) \right)$$

$$= \frac{p \cdot V \cdot \sin(\delta)}{L_q \cdot \omega} \cdot \left(\Phi_{exc} \cdot \rho + (L_d - L_q) \cdot \frac{V \cdot \cos(\delta)}{L_d \cdot \omega} \right) \qquad [2.10]$$

From these last expressions for the torque, it would appear that the latter can be controlled by fixing either the armature current vectors or the armature voltage vectors (Lajoie-Mazenc and Viarouge 1991).

2.2.2. Control modes

Two operational or control modes can be identified for synchronous machines (Lajoie-Mazenc and Viarouge 1991):

– open-loop control mode;

– closed-loop control mode.

For the first mode, it is a matter of fixing the operating speed by choosing an appropriate armature supply frequency, with the two being proportional. The machine can be supplied with either a voltage or a current. This operational mode corresponds to scalar control since the imposed frequency and speed quantities are scalars. The direct supply from the grid is a special case of this operational mode, for which the speed is constant (proportional to the frequency of the network).

This type of operation can be unstable (Lajoie-Mazenc and Viarouge 1991; Grellet and Clerc 1997) and is well suited for applications where the load is known and where the machine is sized, or chosen in such a way as to respond, without risk of stalling (Lajoie-Mazenc and Viarouge 1991). These notions of "stability" and "stall" will be explained in the next section.

The second closed-loop operational mode, which represents almost all variable-speed drives (Lajoie-Mazenc and Viarouge 1991), corresponds to vector control, where the machine is said to be "auto-piloted". Here, again, the machine can be supplied with either a voltage or a current. Typically, they are supplied with a current-regulated voltage source (Grellet and Clerc 1997). We will see that it may be advantageous to switch to a voltage-regulated power supply, in order to extend the operational speed range.

This switch to a voltage-regulated power supply may be necessary for some synchronous machines to optimize their operation, while still remaining in the so-called "*brushless AC*" operation or, by necessity, when switching from a "*brushless AC*" to a "*brushless DC*" operation (Miller 1989; Lajoie-Mazenc and Viarouge 1991; Grellet and Clerc 1997; Louis and Bergman 1999; Shi et al. 2006; Zhu et al. 2006; Miyamasu and Akatsu 2011; Nam 2019). This change is often chosen when it is not possible to regulate the current due to a high EMF value, which itself is linked to an increase in speed (Miller 1989; Lajoie-Mazenc and Viarouge 1991; Grellet and Clerc 1997; Louis and Bergman 1999; Shi et al. 2006; Zhu et al. 2006; Miyamasu and Akatsu 2011; Nam 2019). In this case, we switch to a full-wave voltage supply, "*brushless DC*".

For vector-controlled auto-piloted synchronous machines, it is necessary to control the armature quantities (amplitudes, frequency and phases of the armature voltages and currents) with respect to the position and desired speed of the rotor, as well as the desired value of the torque (Lajoie-Mazenc and Viarouge 1991).

The control laws that we study in this book are developed within the framework of vector-controlled self-piloted synchronous machines, i.e. in the "*brushless AC*" operational mode, to remain consistent with the first harmonic hypothesis.

2.3. Torque characteristics and basic control laws

Torque characteristics are essential when defining control laws and studying them will allow us to appreciate the limiting torque production capacities of different synchronous machines (envelopes). The lossless model is used for this purpose, though we will not completely ignore the notion of efficiency.

In terms of equation [2.10], it is possible to analyze the torque characteristics by studying its variations as a function of either the armature current vectors (I and ψ), or the armature voltage vectors (V and δ). Historically, before the advent of power electronics, electrical machines were coupled directly to electrical networks, which forced their voltage and frequency values to set values, and the torque characteristics then were studied as functions of the armature voltages (V and δ) (Barret 1978; Lajoie-Mazenc and Viarouge 1991). Although both approaches are discussed, in what follows, the torque analysis as a function of the armature voltage is the one most often used.

Besides the historical aspects, two other fundamental reasons can be put forth to justify this choice:

1) the direct link between the stability of the open-loop operation with the amplitude of the armature voltage V and the internal angle δ;

2) to simplify the study of the operation during temporary overload.

We may study the torque characteristics by distinguishing between machines with non-salient poles ($\rho = 1$) and those with salient poles ($\rho \neq 1$). This study is a continuation and an extension of similar studies (Schiferl and Lipo 1990; Lajoie-Mazenc and Viarouge 1991; Soong and Miller 1994; Multon et al. 1995; Soulard 1998; Soulard and Multon 1999; Amara 2001; Amara et al. 2004a; De Doncker et al. 2011; Nam 2019) and takes all types of synchronous machines into account. The main steps allowing for the derivation of important results are presented, but the long mathematical developments are not stated in detail. Instead, they have been included in the form of a downloadable appendix (https://1drv.ms/b/s!AogSAGtYvycUkHI bc3FbsPoSQjJt?e=U9p85C).

2.3.1. Torque characteristics as a function of I and ψ

In this section, we will study the function $\Gamma_{em}(I, \psi)$ [2.10]. The first step is to solve the following equation:

$$\frac{\partial \Gamma_{em}}{\partial \psi} = 0 \Leftrightarrow 2 \cdot (L_d - L_q) \cdot I \cdot \sin^2(\psi) - \Phi_{exc} \cdot \sin(\psi) - (L_d - L_q) \cdot I = 0$$

When studying this function, we will distinguish between machines with non-salient poles ($\rho = 1$) and those with salient poles ($\rho \neq 1$).

2.3.1.1. Machines with non-salient poles ($\rho = 1$)

For machines with non-salient poles, the previous equation has two solutions corresponding to:

$$\sin(\psi) = 0 \qquad [2.11]$$

One solution corresponds to motor operation ($\psi = 0$), and the other corresponds to generator operation ($\psi = \pi$). For a given value of I, the optimal angle maximizing the motor torque is given by:

$$\psi_{opt} = 0 \qquad [2.12]$$

For this angle, the motor torque is then given by:

$$\Gamma_{em}(\psi_{opt}) = p \cdot \Phi_{exc} \cdot I$$

This torque is maximal at the maximum magnitude of the armature current I_{max} and the maximum field flux $\Phi_{exc\ max}$.

2.3.1.2. *Machines with salient poles* ($p \neq 1$)

For machines with salient poles, the following change of variable: $x = \sin(\psi)$ gives rise to a quadratic equation in x, whose discriminant is always positive:

$$\Delta = \Phi_{exc}^2 + 8 \cdot (L_d - L_q)^2 \cdot I^2$$

Two solutions are then possible:

$$\begin{cases} x_1 = \dfrac{\Phi_{exc} - \sqrt{\Delta}}{4 \cdot (L_d - L_q) \cdot I} \\ x_2 = \dfrac{\Phi_{exc} + \sqrt{\Delta}}{4 \cdot (L_d - L_q) \cdot I} \end{cases}$$

With the adopted change of variable, we must verify that $|x| \leq 1$. We may show that:

$$\begin{cases} |x_1| \leq 1 \\ |x_2| \leq 1 \end{cases} \Leftrightarrow \dfrac{|L_d - L_q| \cdot I}{\Phi_{exc}} \geq 1$$

When both solutions satisfy this condition, we have: $|\Gamma_{em}(x_1)| > |\Gamma_{em}(x_2)|$. For a given value of I, the optimal angle that maximizes the motor torque is then given by:

$$\psi_{opt} = \arcsin\left(\dfrac{\Phi_{exc} - \sqrt{\Delta}}{4 \cdot (L_d - L_q) \cdot I}\right), \; \Gamma_{em}(I, \psi_{opt}) > 0 \qquad [2.13]$$

Just as for the machines with non-salient poles, we see that the torque is maximal when the armature current amplitude and the field flux attain their maximal values.

2.3.2. Torque characteristics as a function of V and δ

In this section, we will study the function $\Gamma_{em}(V, \delta)$ [2.9]. The first step is to solve the following equation:

$$\frac{\partial \Gamma_{em}}{\partial \delta} = 0 \Leftrightarrow (1-\rho)\cdot\left(\frac{V}{\omega}\right)\cdot 2\cdot\cos^2(\delta) + \Phi_{exc}\cdot\rho\cdot\cos(\delta) - (1-\rho)\cdot\left(\frac{V}{\omega}\right) = 0$$

Again, when studying the function, we will distinguish between machines with non-salient poles ($\rho = 1$) and those with salient poles ($\rho \neq 1$).

2.3.2.1. Machines with non-salient poles (ρ = 1)

For machines with non-salient poles, the previous equation has two solutions corresponding to:

$$\cos(\delta) = 0 \qquad [2.14]$$

One solution corresponds to motor operation ($\delta = \pi/2$), whereas the other corresponds to generator operation ($\psi = -\pi/2$). For a given value of V, the optimal angle maximizing the motor torque is then given by:

$$\delta_{opt} = \frac{\pi}{2} \qquad [2.15]$$

For this angle, the motor torque is given by:

$$\Gamma_{em}(\delta_{opt}) = \frac{\Phi_{exc}\cdot V}{\Omega\cdot L_d}$$

This torque is maximal when the magnitude of the armature voltage V_{max} and the field flux attain their maximal values. It should be noted that the current amplitude [2.7] under these conditions can be greater than the nominal current.

2.3.2.2. Machines with salient poles (ρ ≠ 1)

For machines with salient poles, the following change of variable: $x = \cos(\delta)$ leads to a quadratic equation in x, whose discriminant is always positive:

$$\Delta = (\Phi_{exc} \cdot \rho)^2 + 8 \cdot (1-\rho)^2 \cdot (V/\omega)^2$$

Two solutions are then possible:

$$\begin{cases} x_1 = \dfrac{-\Phi_{exc} \cdot \rho - \sqrt{\Delta}}{4 \cdot (1-\rho) \cdot (V/\omega)} \\ x_2 = \dfrac{-\Phi_{exc} \cdot \rho + \sqrt{\Delta}}{4 \cdot (1-\rho) \cdot (V/\omega)} \end{cases}$$

For the adopted change of variable, we must verify that $|x| \leq 1$, and it can be shown that:

$$\begin{cases} |x_1| \leq 1 \\ |x_2| \leq 1 \end{cases} \Leftrightarrow \dfrac{|1-\rho| \cdot V}{\Phi_{exc} \cdot \rho \cdot \omega} \geq 1$$

When both solutions satisfy this condition, we have that: $|\Gamma_{em}(x_2)| > |\Gamma_{em}(x_1)|$. For a given value of V, the optimal angle maximizing the motor torque is then given by:

$$\delta_{opt} = \arccos\left(\dfrac{-\Phi_{exc} \cdot \rho + \sqrt{\Delta}}{4 \cdot (1-\rho) \cdot (V/\omega)}\right), \ \Gamma_{em}(V, \delta_{opt}) > 0 \quad [2.16]$$

As for the machines with non-salient poles, it can be shown that the torque is maximal when the armature voltage amplitude and the field flux both attain their maximal values. Also, the current amplitude [2.7] under these conditions can be greater than the nominal current.

2.3.3. *Notion of stability for an open loop and the consequences of closed-loop operations*

A synchronous machine operating in an open loop will have a higher risk of instability (Barret 1978; Lajoie-Mazenc and Viarouge 1991). For a synchronous machine being supplied by a voltage of amplitude V, its stability corresponds to the condition that (Barret 1978):

$$\partial \Gamma_{em}/\partial \delta > 0 \qquad [2.17]$$

We may show that, in the optimal motor operating range, the "armature voltage" vector is always ahead of the "EMF" vector. An increase in the torque load (the resistive torque) necessarily induces the machine to slow down and consequently causes the "armature voltage"/"EMF" phase shift to increase. If condition [2.17] is met, the motor torque allows for the machine to compensate for this increase and thus maintain operational stability. If instead we move into the zone where $\partial \Gamma_{em}/\partial \delta < 0$, then the machine will stall (Barret 1978).

In the following, we will analyze the stability of non-salient pole and salient pole synchronous machines. This study also allows us to see the differences compared to closed-loop operations, for which the instability problem is solved using controls. This will allow us to prove the theorem stated below.

THEOREM.– *For a given torque load and excitation flux, at least two combinations (V, δ), and at most four, will yield the corresponding motor torque. The minimum armature current magnitude corresponds to an operation in the zone where $\partial \Gamma_{em}/\partial \delta > 0$, for δ∈ [0, π].*

2.3.3.1. Machines with non-salient poles (ρ = 1)

For machines with non-salient poles, motor operation corresponds to an internal angle of:

$$\delta \in [0, \pi] \qquad [2.18]$$

The stable operation corresponds to:

$$\delta \in [0, \pi/2] \qquad [2.19]$$

From expression [2.7], we see that the current increases as a function of the internal angle when in the motor operating region. Moreover, for a given torque load, two operating points are possible (Figure 2.5). The first "A" is in the stable zone, whereas the second "B" is in the unstable zone. It follows that it is more interesting to operate in the so-called stable zone, even for closed loops, since the current amplitude is lower and induces fewer armature Joule losses.

2.3.3.2. Machines with salient poles ($\rho \neq 1$)

Depending on the values of the parameters used in the previous model (section 2.2.1.2), for salient pole machines, there are two different forms of torque characteristic as a function of the internal angle. Table 2.3 summarizes the two cases.

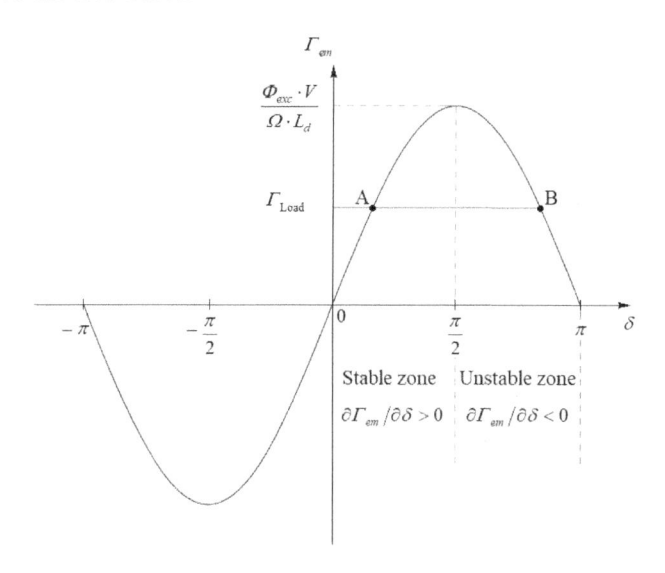

Figure 2.5. *Torque against internal angle δ characteristic for non-salient pole machines*

Parameters	Features $\Gamma_{em}(\delta)$
$\left(\dfrac{\|1-\rho\| \cdot V}{\Phi_{exc} \cdot \rho \cdot \omega}\right) < 1$	The characteristics $\Gamma_{em}(\delta)$ have two optimal values
$\left(\dfrac{\|1-\rho\| \cdot V}{\Phi_{exc} \cdot \rho \cdot \omega}\right) \geq 1$	The characteristics $\Gamma_{em}(\delta)$ have four optimal values

Table 2.3. *Torque characteristics as a function of the internal angle δ for salient pole machines*

Both cases are considered in the following two sections 2.3.3.2.1 and 2.3.3.2.2.

2.3.3.2.1. Case $\left(\dfrac{|1-\rho|\cdot V}{\Phi_{exc}\cdot \rho\cdot \omega}\right)<1$

For these machines, motor operation corresponds to an internal angle within the range of:

$$\delta\in[0,\ \pi]\qquad [2.20]$$

The stable operation corresponds to:

$$\delta\in\left[0,\ \delta_{opt}\right]\qquad [2.21]$$

The behavior of the current amplitude [2.7] depends on the machine parameters. Table 2.4 summarizes the different cases.

The angle δ_0, presented in Figure 2.6, is expressed by:

$$\delta_0 = \arccos\left(\left(\dfrac{\rho^2}{\rho^2-1}\right)\cdot\dfrac{\Phi_{exc}\cdot\omega}{V}\right)\in[0,\ \pi]$$

It can be shown that operating in the so-called stable zone is often more interesting, even for closed-loop operations, since the current amplitude is lower, inducing fewer joule losses in the armature. The details of these proofs can be found in the downloadable appendix (https://1drv.ms/b/s!AogSAGtYvycUkHIbc3FbsPoSQjJt?e=U9p85C).

Parameters	Features $I(\delta)$
$\left(\dfrac{\left\|1-\rho^2\right\|\cdot V}{\Phi_{exc}\cdot\rho^2\cdot\omega}\right)\leq 1$	$\partial I/\partial\delta > 0$
$\left(\dfrac{\left\|1-\rho^2\right\|\cdot V}{\Phi_{exc}\cdot\rho^2\cdot\omega}\right)>1$	– For $\rho<1$, the sign of $\partial I/\partial\delta$ is given by Figure 2.6(a) – For $\rho>1$, the sign of $\partial I/\partial\delta$ is given by Figure 2.6(b)

Table 2.4. *I(δ) characteristics for salient pole machines*

```
0   +      δ_opt    +   π/2   +    δ_0     —        π
├──────────┼──────────┼──────────┼──────────┤

                    a) ρ < 1

0    —      δ_0     +   π/2   +   δ_opt    +        π
├──────────┼──────────┼──────────┼──────────┤

                    b) ρ > 1
```

Figure 2.6. *The sign of $\partial I/\partial \delta$ for a few machines with salient poles (Table 2.4)*

2.3.3.2.2. Case $\left(\dfrac{|1-\rho| \cdot V}{\Phi_{exc} \cdot \rho \cdot \omega} \right) \geq 1$

In this case, the torque characteristics as a function of the internal angle have four optimal values. Figure 2.7 shows the variation of the sign of $\partial \Gamma_{em}/\partial \delta$.

The angle δ_2, shown in Figure 2.7, is expressed as:

$$\delta_2 = \arccos\left(\dfrac{-\Phi_{exc} \cdot \rho - \sqrt{(\Phi_{exc} \cdot \rho)^2 + 8 \cdot (1-\rho)^2 \cdot (V/\omega)^2}}{4 \cdot (1-\rho) \cdot (V/\omega)} \right) \in [0,\ \pi]$$

For these machines, the motor operation corresponds to an internal angle within the ranges of:

$$\begin{cases} \rho < 1 \Rightarrow \delta \in [-\pi,\ -\delta_1] \cup [0,\ \delta_1] \\ \rho > 1 \Rightarrow \delta \in [-\delta_1,\ 0] \cup [\delta_1,\ \pi] \end{cases} \qquad [2.22]$$

The stable operation corresponds to:

$$\begin{cases} \rho < 1 \Rightarrow \delta \in [-\pi,\ -\delta_2] \cup [0,\ \delta_{opt}] \\ \rho > 1 \Rightarrow \delta \in [-\delta_1,\ -\delta_2] \cup [\delta_1,\ \delta_{opt}] \end{cases} \qquad [2.23]$$

The angle δ_1, at which the electromagnetic torque $\Gamma_{em}(\delta)$ vanishes, can be expressed by:

$$\delta_1 = \arccos\left(\left(\frac{\rho}{\rho-1}\right) \cdot \frac{\Phi_{exc} \cdot \omega}{V}\right) \in [0, \pi]$$

For this type of machine, the current amplitude [2.7], behaves identically to that shown in Figure 2.6.

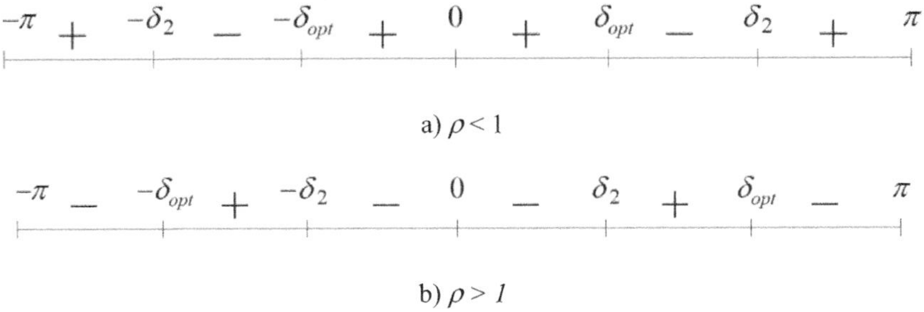

$-\pi \quad + \quad -\delta_2 \quad - \quad -\delta_{opt} \quad + \quad 0 \quad + \quad \delta_{opt} \quad - \quad \delta_2 \quad + \quad \pi$

a) $\rho < 1$

$-\pi \quad - \quad -\delta_{opt} \quad + \quad -\delta_2 \quad - \quad 0 \quad - \quad \delta_2 \quad + \quad \delta_{opt} \quad - \quad \pi$

b) $\rho > 1$

Figure 2.7. Sign of $\partial \Gamma_{em}/\partial \delta$

It can be shown that the operation allowing for the optimization of the armature joule losses corresponds to the intervals:

$$\begin{cases} \rho < 1 \Rightarrow \delta \in [0, \delta_{opt}] \\ \rho > 1 \Rightarrow \delta \in [\delta_1, \delta_{opt}] \end{cases}$$

As before, details of these proofs can be found in the downloadable appendix.

2.3.3.3. Synthesis

The reader is strongly advised to consult the downloadable appendix (https://1drv.ms/b/s!AogSAGtYvycUkHIbc3FbsPoSQjJt?e=U9p85C). This document contains details of the mathematical proofs and developments. The latter is carried out with an emphasis on the formal analytical approach and, when a numerical approach is required, "Matlab" scripts are provided.

These developments conclude that the theorem stated at the beginning of section 2.3.3 is indeed valid. It follows that it is more attractive to operate in the regions where $\partial \Gamma_{em}/\partial \delta > 0$, with $\delta \in [0, \pi]$, even for closed-loop controlled machines, for which stability is ensured by their control.

2.3.4. *Fundamental control laws*

Depending on the type of synchronous machines (fixed or controllable field flux), and the type of power supply adopted (voltage or current supply), several control strategies can be implemented.

There are different reasons why control strategies are studied. If a machine has already been built, then studying the control laws can contribute towards imposing certain quantities, such as the torque, or a unitary power factor, to adapt the machine for its intended applications (Lajoie-Mazenc and Viarouge 1991), while optimizing its efficiency.

Studies can also be conducted to integrate the control aspects from the design stage of the machine. It is interesting to understand which parameter sets are sufficient for the machine to optimally perform its intended application. The machine is then no longer considered as an isolated component, but as an element interacting with the rest of its environment (Lajoie-Mazenc and Viarouge 1991).

2.3.4.1. *Operation at maximum torque per ampere (MTPA)*

For a given armature current amplitude and field flux, the "maximum torque per ampere control" (MTPA control) consists of fixing the angle ψ which maximizes the torque. This angle is given by equations [2.12] and [2.13], for machines with non-salient and salient poles, respectively. Note that this angle does not depend on the speed.

The maximum torque is obtained when the armature current amplitude and the field flux are maximal. Depending on the type of synchronous machine, we may fix either the values of the armature current amplitude and the angle ψ_{opt} (i.e. machines with a fixed field flux), or the values of the armature current amplitude, the field flux and the angle ψ_{opt} (i.e. machines with controllable field flux). The latter type of machine allows for greater flexibility.

For salient pole machines, the angle ψ_{opt} corresponds to a magnetizing operation for $\rho < 1$, i.e. the armature magnetic reaction field is additive to the field flux along the direct axis, and a demagnetizing operation for $\rho > 1$, i.e. the armature magnetic reaction opposes the field flux along the direct axis.

This type of control can be implemented directly using the current supply to the machine, or more frequently by using a current-regulated voltage supply (Grellet and Clerc 1997). Controlling the current amplitude ensures that any thermal constraints are always met (Nam 2019).

2.3.4.2. Maximum torque per volt (MTPV) operation

This control is dual to that of the previous control. For a given armature voltage amplitude, it consists of fixing the angle δ maximizing the torque for a given field flux and speed. This angle is determined by equations [2.15] and [2.16] for machines with non-salient poles and salient poles, respectively.

The maximum torque is obtained using the maximal armature voltage amplitude and field flux values. When applying this control law, it is necessary to verify that the current amplitude does not exceed its nominal value, by allowing any thermal constraints to be respected. This is particularly necessary during low-speed (low-frequency) operation, in which the equivalent impedance of the machine is relatively low, even if the maximum field flux is fixed. It is then important to connect the voltage amplitude to the frequency value (speed) (Lajoie-Mazenc and Viarouge 1991; Grellet and Clerc 1997). This is called "V/f" control.

This control law can be implemented using a voltage inverter power supply that allows for the voltage amplitude, its phase shift with respect to the EMF (internal angle δ) and its frequency to be controlled.

2.3.4.3. Unity power factor (UPF) operation

An operation whose unity power is 1 allows for the converter, which supplies the corresponding machine, to be dimensioned as accurately as possible (Lajoie-Mazenc and Viarouge 1991; Multon et al. 1995).

The power factor is given by:

$$F_p = \cos(\varphi) = \cos(\delta - \psi) = \frac{v_d \cdot i_d + v_q \cdot i_q}{V \cdot I} \qquad [2.24]$$

where φ is the "armature voltage/armature current" phase shift angle.

It can also be expressed as follows:

$$F_p = \left(\frac{(L_q - L_d) \cdot I \cdot \sin(\psi) + \Phi_{exc}}{V}\right) \cdot \omega \cdot \cos(\psi)$$

$$F_p = \left(\frac{L_q \cdot \omega \cdot \Phi_{exc} + (L_d - L_q) \cdot V \cdot \cos(\delta)}{I \cdot \omega \cdot L_d \cdot L_q}\right) \cdot \sin(\delta)$$

To impose the unity power factor, the first step is to solve the following equation:

$$F_p = 1 \Rightarrow (L_q - L_d) \cdot I \cdot \sin^2(\psi) + \Phi_{exc} \cdot \sin(\psi) - L_q \cdot I = 0$$

This function can be studied by distinguishing between machines with non-salient poles ($\rho = 1$) and those with salient poles ($\rho \neq 1$).

2.3.4.3.1. Machines with non-salient poles ($\rho = 1$)

For machines with non-salient poles, the previous equation has two solutions that correspond to:

$$\sin(\psi) = \frac{L_d \cdot I}{\Phi_{exc}} \qquad [2.25]$$

One corresponds to motor operation ($\cos(\psi) > 0$), and the other to generator operation ($\cos(\psi) < 0$).

As the angle in the solution that corresponds to motor operation is different from the optimal angle that maximizes the torque [2.12], we deduce that the torque is not maximized in adopting this control law (Lajoie-Mazenc and Viarouge 1991).

2.3.4.3.2. Machines with salient poles ($\rho \neq 1$)

For machines with salient poles, the following change of variable: $x = \sin(\psi)$ gives rise to a quadratic equation in x, whose discriminant is:

$$\Delta = \Phi_{exc}^2 \cdot \left(1 + 4 \cdot (\rho - 1) \cdot \rho \cdot \frac{L_d^2 \cdot I^2}{\Phi_{exc}^2}\right)$$

Figure 2.8.a presents how the sign of the discriminant varies within the plane with coordinates $\left(\frac{L_d \cdot I}{\Phi_{exc}}, \rho\right)$. The boundary corresponding to $\Delta = 0$ is given by:

$$\left(\rho < 1 \text{ and } \frac{L_d \cdot I}{\Phi_{exc}} = \frac{1}{2 \cdot \sqrt{(1-\rho) \cdot \rho}}\right)$$

In the case where $\Delta \geq 0$, we have two solutions:

$$\begin{cases} x_1 = \dfrac{\Phi_{exc} - \sqrt{\Delta}}{2 \cdot (L_d - L_q) \cdot I} \\[2mm] x_2 = \dfrac{\Phi_{exc} + \sqrt{\Delta}}{2 \cdot (L_d - L_q) \cdot I} \end{cases}$$

In terms of the adopted change of variable, it is necessary to check that $|x| \leq 1$. It can be shown that:

$$\begin{cases} |x_1| \leq 1 \Leftrightarrow \begin{cases} \left[\rho > \dfrac{1}{2} \text{ and } \dfrac{L_d^2 \cdot I^2}{\Phi_{exc}^2} \leq 1\right] \\[2mm] \left[\rho \leq \dfrac{1}{2} \text{ and } \dfrac{L_d^2 \cdot I^2}{\Phi_{exc}^2} < \dfrac{1}{4 \cdot (1-\rho) \cdot \rho}\right] \end{cases} \\[6mm] |x_2| \leq 1 \Leftrightarrow \left[\rho \leq \dfrac{1}{2} \text{ and } 1 \leq \dfrac{L_d^2 \cdot I^2}{\Phi_{exc}^2} < \dfrac{1}{4 \cdot (1-\rho) \cdot \rho}\right] \end{cases}$$

For solutions satisfying the condition $|x| \leq 1$, the angle ensuring a unity power factor is then given by:

$$\psi_{UPF} = \begin{cases} \arcsin(x_1) \\ \arcsin(x_2) \end{cases}, \quad \Gamma_{em}(I, \psi_{UPF}) > 0 \qquad [2.26]$$

As for the non-salient pole machines, it can be shown that the torque is lower when this control law is used, compared to when the "MTPA" control law is used.

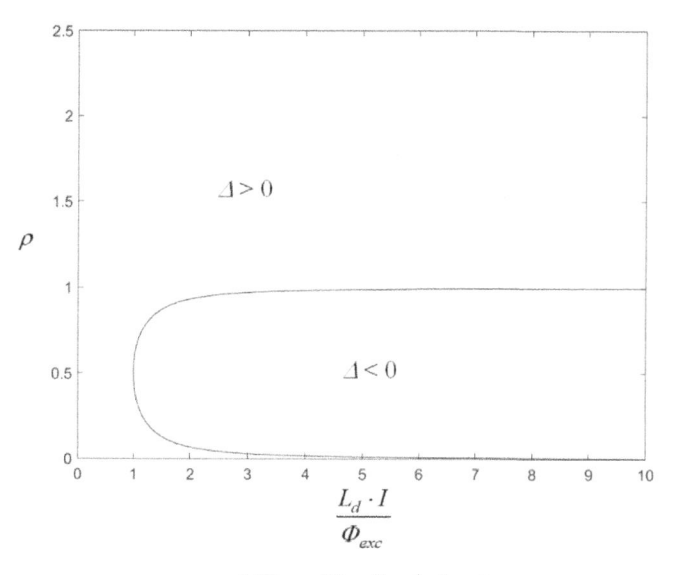

Figure 2.8. *Unity power factor operation*

a) Sign of the discriminant

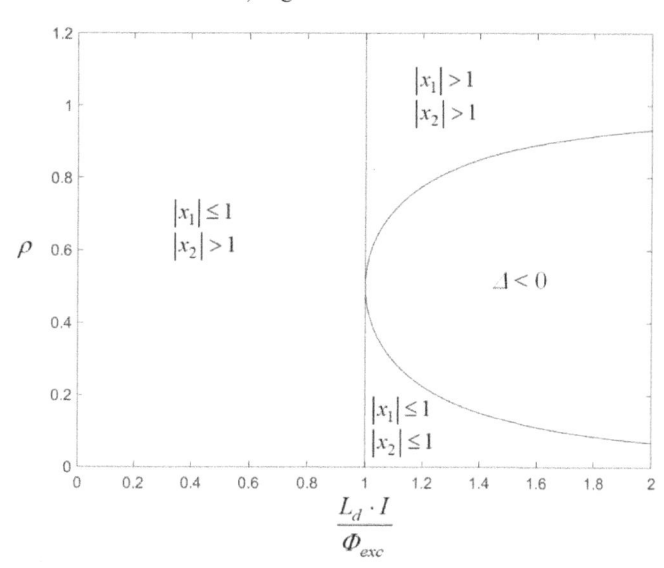

b) Existence of the solutions

2.3.5. *Temporary overloaded motor operation*

Overloading the drive motor operation is sometimes necessary to enable vehicles to pass exceptional obstacles. The overload capability is studied in this section only for low speeds because, at high speeds, the overload capability is then used for comfort while driving, such as when on a highway, whereas for low speeds, it is used to maintain operational continuity instead, such as when climbing up a steep hill.

Overloaded operations often require the ability to convert more power than that which it is rated for, and also to impose a torque greater than the rated torque for a limited period of time. The purpose of limiting the overloaded operating time is to ensure that the thermal constraints are met, while preventing the drive motor from deteriorating more quickly.

Temporary overloaded operations may be achieved within all the constraints by oversizing the traction motor. The nominal operation would then correspond to a degraded operational mode. This is called exceptional operation, rather than overloaded operation, and is not generally the adopted choice. The other option (temporary overloaded operation) is to prevent the drive system from meeting the current or voltage constraints, or both simultaneously, thus temporarily allowing more power to be converted. Since the operating time under these conditions is limited, no damage to the drive system is expected.

Overloaded operations always require the armature current magnitude to be greater than the rated current, regardless of the drive motor parameters. In Figure 2.9(a), we have represented the boundary between two zones when the torque overload is equal to twice the nominal torque.

Zone 1 corresponds to the parameters $\left(L_d \cdot I_{max}/\Phi_{exc\,max}, \rho\right)$ for which this torque gain can only be achieved by using a greater current than the maximal current ($I > I_{max}$) (the voltage constraint can be met ($V \leq V_{max}$)). Zone 2 corresponds to the parameters for which the overloaded torque can only be achieved with a current greater than the maximum current and with a voltage greater than the maximum voltage. The boundary corresponds to the parameters for which the maximum torque, by adopting the MTPV control law, is equal to twice that obtained from imposing the MTPA control law.

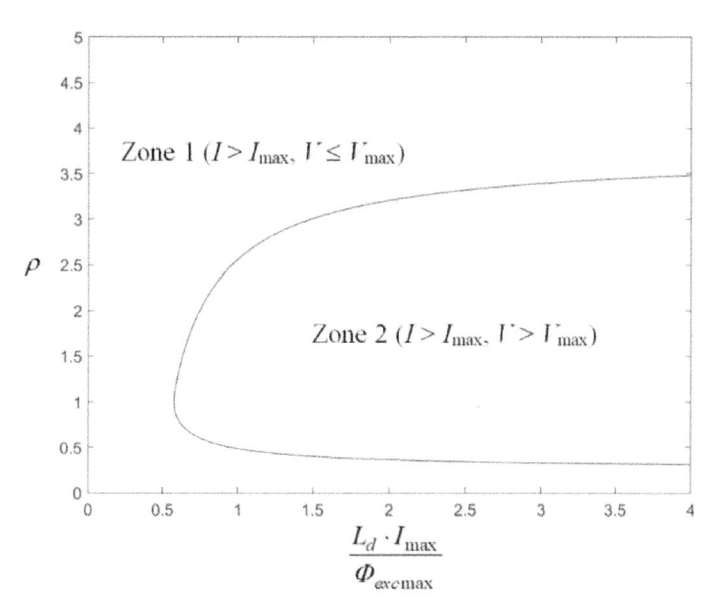

a) An overloaded operation with the maximal torque twice the nominal value

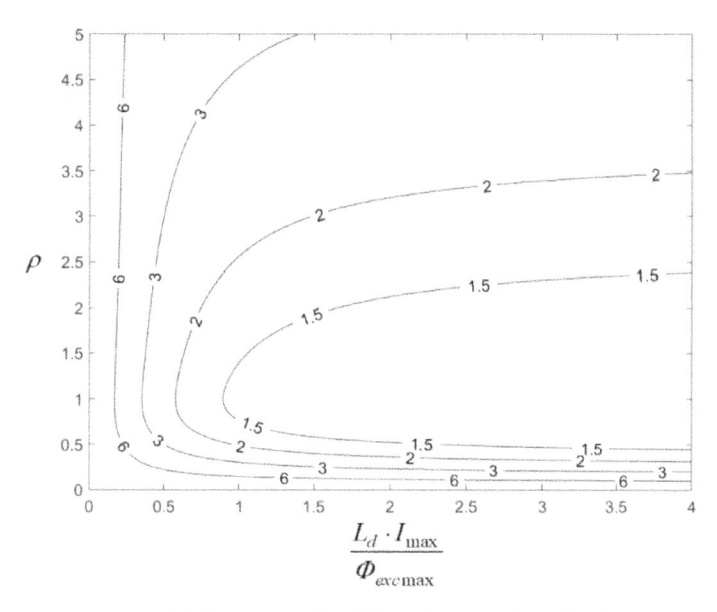

b) Boundaries for different levels of overload

Figure 2.9. *Influence that overloading has on the current and voltage parameters*

Figure 2.9(b) presents the boundaries between different overloads with respect to the nominal torque in the plane $\left(L_d \cdot I_{max}/\Phi_{exc\,max}, \rho\right)$. It can be said that the lower the values of $\left(L_d \cdot I_{max}/\Phi_{exc\,max}, \rho\right)$ are, the less constraining the overload on the converter–machine assembly sizing.

2.4. Setting the speed of HESMs (maximal characteristics/envelopes)

Variable-speed drives are used in many industrial and domestic applications. In order to optimize the efficiency with varying speeds, it is necessary to maximize the torque produced for a given input power. The full-load operation (envelopes) of synchronous motors is the subject of this section.

After reviewing the fundamental control laws introduced in the previous section, we will consider how to combine them in the context of variable-speed applications. We recall that electric traction/propulsion was chosen as the typical application. For a given speed, the aim is to determine the operating conditions that maximize the useful torque delivered by a motor for a given set of parameters. The optimal control strategies will be defined for all possible parameter sets.

Since full-load operations are not the most highly stressed operation in the standard operating cycle of motor vehicles (Amara 2001; Amara et al. 2004a), we will not be particularly interested in studying the efficiency of this operation, even though maximizing the torque for a given input power implicitly means maximizing it. Therefore, we will neglect any losses in this study. The study of this operation over the entire "torque/speed" plane will be the subject of the following section, which also takes into account any losses.

To present the benefits obtained when the excitation flux is controlled, compared to just having a constant excitation flux, the control strategies to be adopted are studied in each of the two cases.

2.4.1. Low-speed operations

At low speeds, synchronous motors have relatively low impedances, so it is possible to set the maximal current with a relatively low voltage. To impose the maximal torque, we will use the MTPA control law while fixing the maximal current I_{max} (nominal current) and the maximal excitation flux $\Phi_{exc\,max}$ [2.27].

$$\begin{cases} I = I_{max} \text{ and } \Phi_{exc} = \Phi_{exc\,max} \\ \psi_{opt} = \begin{cases} 0 & \rho = 1 \\ \arcsin\left(\dfrac{\Phi_{exc} - \sqrt{\Phi_{exc}^2 + 8 \cdot (L_d - L_q)^2 \cdot I^2}}{4 \cdot (L_d - L_q) \cdot I}\right), \Gamma_{em}(I, \psi_{opt}) > 0 & \rho \neq 1 \end{cases} \\ V = \omega \cdot \sqrt{\left(L_q \cdot I \cdot \cos(\psi_{opt})\right)^2 + \left(\Phi_{exc} - L_d \cdot I \cdot \sin(\psi_{opt})\right)^2} \\ PF = \left(\dfrac{(L_q - L_d) \cdot I \cdot \sin(\psi_{opt}) + \Phi_{exc}}{\sqrt{\left(L_q \cdot I \cdot \cos(\psi_{opt})\right)^2 + \left(\Phi_{exc} - L_d \cdot I \cdot \sin(\psi_{opt})\right)^2}}\right) \cdot \cos(\psi_{opt}) \end{cases}$$

[2.27]

The voltage amplitude is therefore proportional to the speed, and the power factor is therefore independent of the speed. Since the maximal voltage is limited, the maximum speed that can be reached will also be limited when adopting the MTPA control law.

Figure 2.10 shows the variation in the optimal angle ψ_{opt} and the power factor in the plane $(L_d \cdot I_{max} / \Phi_{exc\,max}, \rho)$.

For machines with $\rho < 1$, the armature magnetic reaction is magnetizing (i.e. the total flux is greater than the excited flux along the direct axis), while on the other hand, it is demagnetizing (i.e. the total flux is less than the excited flux along the direct axis) when $\rho > 1$. The power factor is greater when the parameters $(L_d \cdot I_{max} / \Phi_{exc\,max}, \rho)$ are lower, which means that the lower the parameters $(L_d \cdot I_{max} / \Phi_{exc\,max}, \rho)$, the greater the amount of power converted and the better the sizing of the converter-machine assembly (Jahns 2000).

Indeed, from Figure 2.10(b), we note that machines with a salience ratio of $\rho < 1$ have better power factors for a given value of $L_d \cdot I_{max} / \Phi_{exc\,max}$, in particular when the latter is less than 1. This allows for an optimal dimensioning of the converter-machine assembly (Multon et al. 1995; Bianchi et al. 2000; Amara 2001). These machines also have easier high-speed operations (Multon et al. 1995; Bianchi and Bolognani 1997; Bianchi et al. 2000).

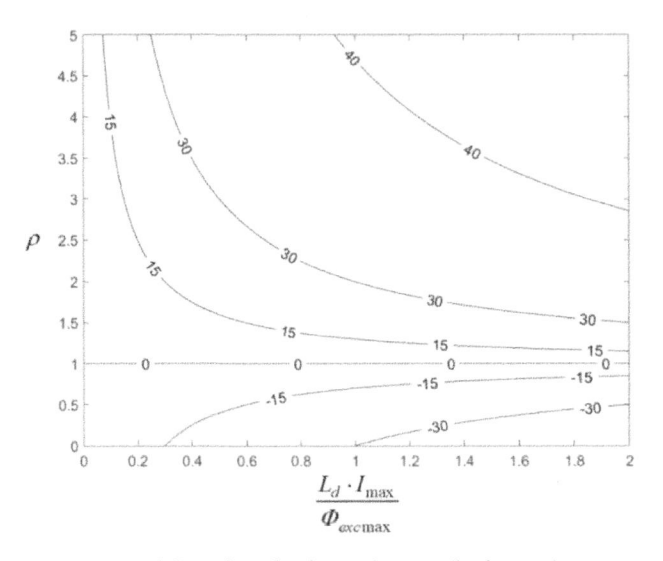

a) Isovalues for the angles ψ_{opt} (in degrees)

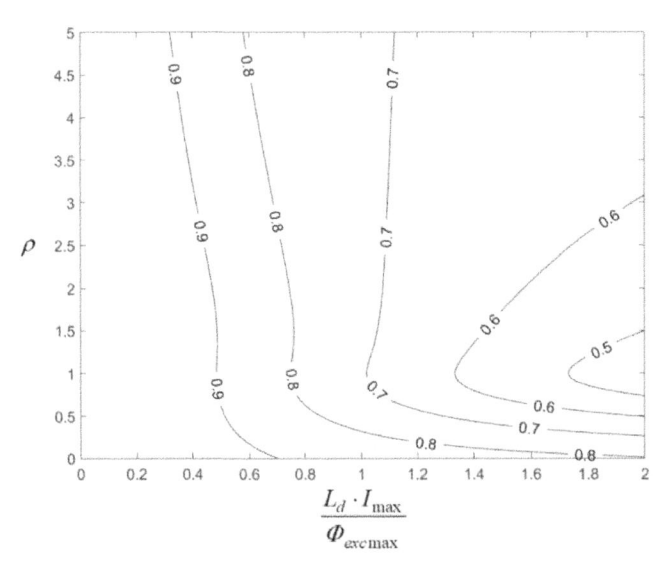

b) Isovalues for the power factors

Figure 2.10. *Operational characteristics when the MTPA control law is adopted*

Machines with magnets embedded into the rotor often have a saliency ratio of $\rho > 1$ (Figure 2.11). To get a saliency ratio of $\rho < 1$, special sizing is required. Figures 2.12(a) and (b) show some solutions which have been proposed in the scientific literature (Ionel et al. 1995, 1996; Chalmers et al. 1996). Figure 2.12(c) shows a solution that includes hybrid excitation.

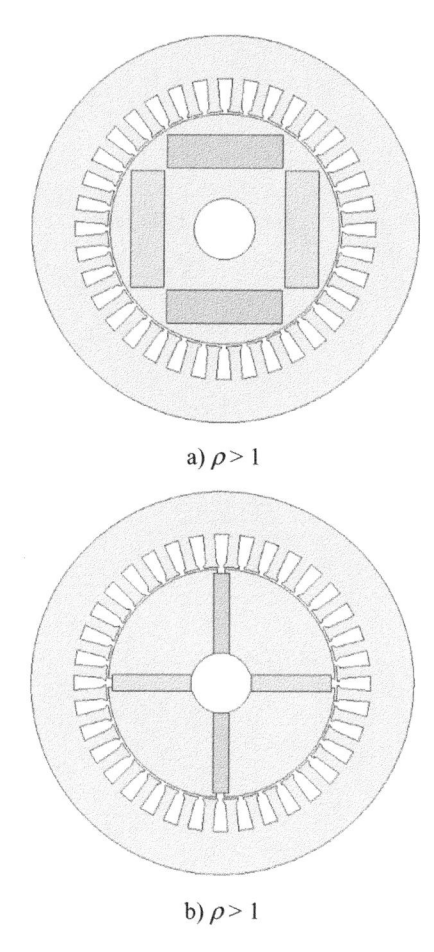

a) $\rho > 1$

b) $\rho > 1$

Figure 2.11. *Permanent magnet synchronous machines with $\rho > 1$*

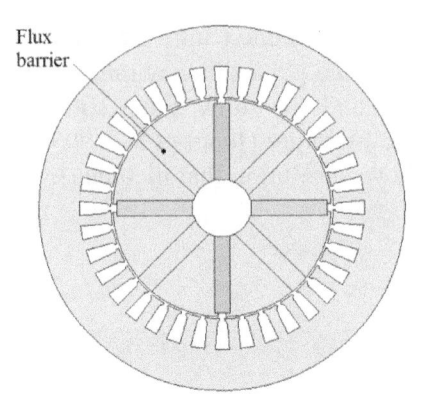

a) $\rho < 1$ (Ionel et al. 1995, 1996)

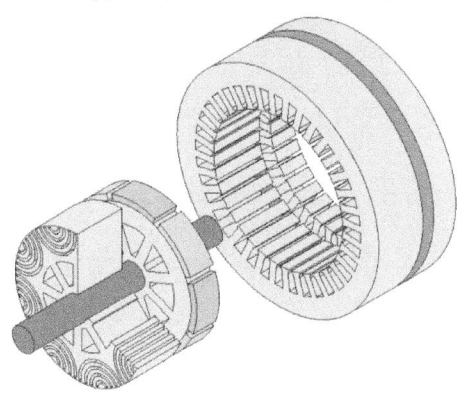

b) $\rho < 1$ (Chalmers et al. 1996)

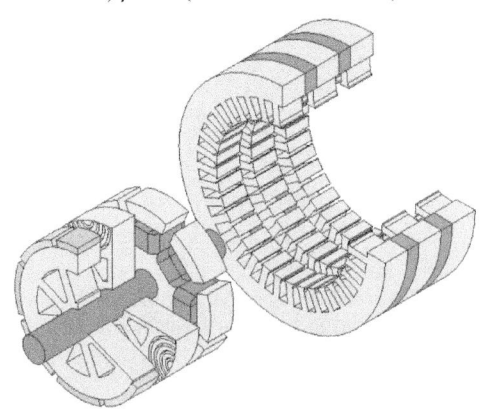

c) Hybrid excited machine with $\rho < 1$

Figure 2.12. *Synchronous machines with $\rho < 1$*

In the rotor of the machine in Figure 2.12(a), the introduction of a flux barrier along the q-axis decreases the inductance along this axis. Depending on the thickness of this barrier, it is possible to obtain a saliency ratio with $\rho < 1$ (Ionel et al. 1995, 1996). Figure 2.12(b) presents another solution (Chalmers et al. 1996), where two rotors, one with surface-mounted permanent magnets ($\rho = 1$), and the other with varying reluctance ($\rho < 1$), are combined within the same stator. This combination results in a machine with $\rho < 1$. The salience ratio can be adjusted by changing the lengths of the two rotors.

The structure shown in Figure 2.21(c) is a hybrid excited machine. This structure has three juxtaposed rotors:

– a surface-mounted permanent magnet rotor ($\rho = 1$);

– a variable-reluctance rotor ($\rho < 1$);

– a wound excitation rotor ($\rho < 1$) (Lajoie-Mazenc and Viarouge 1991).

These three rotors are placed within the same stator. Hybrid excitation is produced by combining the permanent magnet and wound excitation rotors. The variable-reluctance rotor allows for the salience value ratio to be adjusted. As before, it is possible to adjust the parameters of the machine by changing the axial lengths of the three rotors.

2.4.1.1. *The notion of the base speed*

Base speed is the maximum speed that can be reached when adopting the MTPA control law, while also maintaining the armature current and excitation flux amplitudes at their maximal values. In the lossless model, it can be expressed as:

$$\Omega_b = \frac{V_{max}}{p \cdot \sqrt{\left(L_q \cdot I_{max} \cdot \cos(\psi_{opt})\right)^2 + \left(\Phi_{exc\,max} - L_d \cdot I_{max} \cdot \sin(\psi_{opt})\right)^2}} \quad [2.28]$$

This speed delimits two operating zones in particular:

– a maximal torque operational zone ($\Omega \leq \Omega_b$);

– a flux-weakening zone ($\Omega > \Omega_b$).

These two zones are clearly identifiable in Figure 2.4. Beyond the base speed, it is no longer possible to maintain the MTPA control law while still imposing the maxim armature current and excitation flux amplitudes. It is therefore necessary to weaken the flux of the machine (i.e. reducing the flux in the air-gap) to increase the speed. Since the amplitude of the armature supply voltage is limited, given that we know it is proportional to the product of the flux with the speed, it is straightforward to see that an increase in speed is achieved by reducing the flux. The torque is lower in the weakened flux zone (Figure 2.4).

It should be noted that the base speed is not necessarily the nominal speed. The nominal values, which we consider in this book, are those that can be maintained theoretically for an indefinite amount of time, without any risk of the machines aging prematurely.

The idea of the base speed will be used in the following section to define a system of normalized quantities. This system allows for the behavior of synchronous machines to be better understood in general, and in particular for that of hybrid excited machines.

2.4.1.2. *Reduced or normalized values*

In this section, we will define a system of normalized quantities, or reduced parameters, which will clarify the relationship between the performance of synchronous machines and their characteristic parameters (Multon et al. 1995; Schiferl and Lipo 1990; Soong and Miller 1994), thus making it easier to classify them. Normalization allows for dimensionless quantities to be used, which usually reduces the number of influential characteristic parameters. This is the focus of dimensional analysis, which is widely used in fluid mechanics (Harris et al. 1970). The work by Harris et al. (1970) is one of the first contributions of dimensional analysis applied to electric machines.

Without explicitly stating it, we have already used some reduced quantities in the previous sections (Figures 2.8 to 2.10). The first reduced parameter was defined in equation [2.9]: the saliency ratio ρ. We saw that it allowed for machines with non-salient poles ($\rho = 1$) to be distinguished from machines with salient poles ($\rho \neq 1$). The second dimensionless parameter that emerged from our previous analysis was defined as follows:

$$L_{dn} = \frac{L_d \cdot I_{max}}{\Phi_{exc\,max}} \qquad [2.29]$$

This is the reduced direct axis inductance. The plane ($L_d \cdot I_{max}/\Phi_{exc\,max}$, ρ) allowed us to establish interesting classifications which are illustrated in Figures 2.8–2.10. This is the plane that we will use when studying high-speed operations (flux-weakening operations).

Soong and Miller (1994) introduced an alternative design, which they used to analyze flux-weakening operations (high-speed operations) of conventional permanent magnet synchronous machines ($\rho \geq 1$), and variable-reluctance synchronous machines. Of course, they have only considered machines with $\rho \geq 1$. We will extend this study to the case of synchronous machines with $\rho < 1$.

As their study concerned both conventional permanent magnet machines ($\rho \geq 1$) and variable-reluctance synchronous machines, the authors considered a design ($\Phi_{exc\,max}/\Phi_b$, ρ), with Φ_b as the flux when the MTPA control law is adopted, defined in the lossless model by:

$$\Phi_b = \sqrt{\left(L_q \cdot I_{max} \cdot \cos\left(\psi_{opt}\right)\right)^2 + \left(\Phi_{exc\,max} - L_d \cdot I_{max} \cdot \sin\left(\psi_{opt}\right)\right)^2}$$

We have $\Phi_{exc\,max}/\Phi_b \in [0, 1)$, with the zero value corresponding to variable-reluctance machines. In the case of the plane ($L_d \cdot I_{max}/\Phi_{exc\,max}$, ρ), variable-reluctance synchronous machines correspond to the limit $L_{dn} \to +\infty$.

In the context of studying excited synchronous machines (wound excitation synchronous machines, permanent magnet synchronous machines and hybrid excited synchronous machines), the ($L_d \cdot I_{max}/\Phi_{exc\,max}$, ρ) plane is more useful. However, variable-reluctance synchronous machines are also treated in this work, and the following section is devoted to them.

We associate the normalized d–q model (Amara et al. 2009, 2019, 2021a) with the two main parameters ($L_d \cdot I_{max}/\Phi_{exc\,max}$, ρ), and the same normalization system is used for both models, with and without losses.

The normalized values for the armature current and voltage magnitude are defined by:

$$I_n = \frac{I}{I_{max}} \text{ and } V_n = \frac{V}{p \cdot \Omega_b \cdot \Phi_{exc\,max}} \qquad [2.30]$$

The index "n" is used to represent normalized quantities. The reduced values of speed, power and torque are given by:

$$\Omega_n = \frac{\Omega}{\Omega_b}, \quad P_n = \frac{P}{V_{max} \cdot I_{max}} \quad \text{and} \quad \Gamma_n = \frac{P_n}{\Omega_n} \qquad [2.31]$$

For wound excitation synchronous machines, it remains to define the reduced values for the inductance along the q-axis, for the mutual inductance between the wound field excitation circuit and the armature windings, for the excitation flux, as well as for the normalized value of the excitation current:

$$L_{qn} = \rho \cdot L_{dn}, \quad k_{en} = \frac{k_e \cdot I_{e\,max}}{\Phi_{exc\,max}}, \quad k_f = \frac{\Phi_{exc}}{\Phi_{exc\,max}} \quad \text{and} \quad I_{en} = \frac{I_e}{I_{e\,max}} \qquad [2.32]$$

where $I_{e\,max}$ is the maximum excitation current value.

For the model with losses, we must further include normalized values for the quantities and parameters corresponding to the different losses:

$$P_{Cu\,n} = \frac{P_{Cu}}{V_{max} \cdot I_{max}} \text{ (Joule losses)} \quad \text{and} \quad P_{Fe\,n} = \frac{P_{Fe}}{V_{max} \cdot I_{max}} \text{ (iron losses)} \quad [2.33]$$

$$R_{sn} = \frac{R_s \cdot I_{max}}{\rho \cdot \Omega_b \cdot \Phi_{exc\,max}}, \quad R_{fn} = \frac{R_f \cdot I_{max}}{\rho \cdot \Omega_b \cdot \Phi_{exc\,max}} \quad \text{and} \quad R_{en} = \frac{R_e \cdot I_{e\,max}}{V_{e\,max}} \qquad [2.34]$$

where $V_{e\,max}$ is the maximum excitation circuit voltage value.

For wound excitation machines, the ratio between the apparent powers of the converters powering the armature windings and the excitation circuit is also defined:

$$\beta = \frac{V_{max} \cdot I_{max}}{V_{e\,max} \cdot I_{e\,max}} \qquad [2.35]$$

For excited machines (wound excitation machines, permanent magnet machines and hybrid excited machines), the normalized form of the excitation flux equation defined in Figure 2.1(b) is given by:

$$\frac{\Phi_{exc}}{\Phi_{exc\,max}} = \frac{\Phi_a}{\Phi_{exc\,max}} + \frac{k_e \cdot I_e}{\Phi_{exc\,max}} \Rightarrow k_f = \alpha + k_{en} \cdot I_{en} \qquad [2.36]$$

The parameter α is called the hybridization ratio and is of particular importance when we consider dimensional approaches. It is specific to hybrid excited machines, and, in the section dedicated to operations on the entire "torque/speed" plane, we will see how it can be used to optimize the energy efficiency of the drive systems (Amara et al. 2009).

2.4.1.2.1. Variations in the normalized parameter domains

We can classify the various standardized quantities that were introduced above into two main groups: variable electromagnetic and mechanical quantities (I_n, V_n, I_{en}, k_f, P_n, Ω_n, Γ_n), and "constant" parameters.

Table 2.5 provides the ranges over which the different quantities and parameters can vary.

Quantities and parameters	Variation interval
I_n	[0, 1]
V_n	[0, $V_{n\,max}$]
I_{en}	[0, 1]
k_f	[0, 1]
P_n	[0, 1]
Ω_n	[0, +∞)
Γ_n	[0, 1]
ρ	(0, +∞)
L_{dn}	(0, +∞)
R_{sn}	[0, 10]
R_{fn}	(0, +∞)
R_{en}	1
k_{en}	[0, 1]
β	(0, +∞)
α	[0, 1]

Table 2.5. *Variation intervals of the standardized quantities and parameters*

For some quantities or parameters, the indicated ranges may not be completely realistic, in the sense that some values within the range may never be accessible. Having an infinite speed is not realistic and is not necessarily desired. The saliency ratio is often limited. The lowest value found in the scientific literature is 0.1 (Chalmers et al. 1996), and the highest is 20 (Boldea et al. 1992; Soong and Miller 1994). The absence of any losses is also not realistic. However, it is interesting to analyze the operation of the machines from a purely mathematical point of view without setting any physical limits. This is because it is not possible to predict what new techniques or technologies may allow for in the future, such as superconductivity, for example.

Furthermore, it should be noted that, for a given machine, the parameters are considered to be constant. This is also not realistic. Indeed, some physical phenomena, such as magnetic saturation or heating, which depend on the operating conditions, will make these parameters vary. Nevertheless, since the effects of magnetic saturation are understood by designers of electrical machines, the performance analysis on the entire $\left(L_d \cdot I_{max}/\Phi_{exc\,max}, \rho\right)$ plane lets them take this into account indirectly. Heating induces an increase in resistance, which then leads to an increase in Joule losses, but this may also reduce the amount of iron losses. Regarding the latter losses, modeling them with a constant resistance as a function of speed amounts to considering just eddy current losses, although hysteresis losses are also involved. However, since the intervals are large enough, they let us take these variations into account.

Figure 2.13 shows the variation of $V_{n\,max}$ in the (L_{dn}, ρ) plane for the lossless model. This quantity corresponds to the upper limit for the amplitude relative to the "Park" frame for the normalized armature voltage. It can be shown that $V_{n\,max} > 1$ (see the appendix: https://1drv.ms/b/s!AogSAGtYvyc UkHIbc3FbsPoSQjJt?e=U9p85C).

For hybrid excited machines, the lower limit for the excitation coefficient k_f can be greater than zero ($k_{f\,min} > 0$). Indeed, any thermal constraints, magnetic saturation or demagnetization of the permanent magnets may limit the minimum value of this coefficient.

For the normalized armature winding resistance value, an upper limit of 10 was fixed. This is a relatively high value that lets us consider a very large range of machines. Typically, electrical machine designers try to limit the value of this resistance to obtain good efficiencies (Soulard and Multon 1999). Investigating the scientific literature reveals that this value can be relatively high for small machines (Soulard and Multon 1999). Soulard and

Multon (1999) used the same normalized value for this resistance, and they indicate that the highest value they have encountered in the scientific literature is 3.

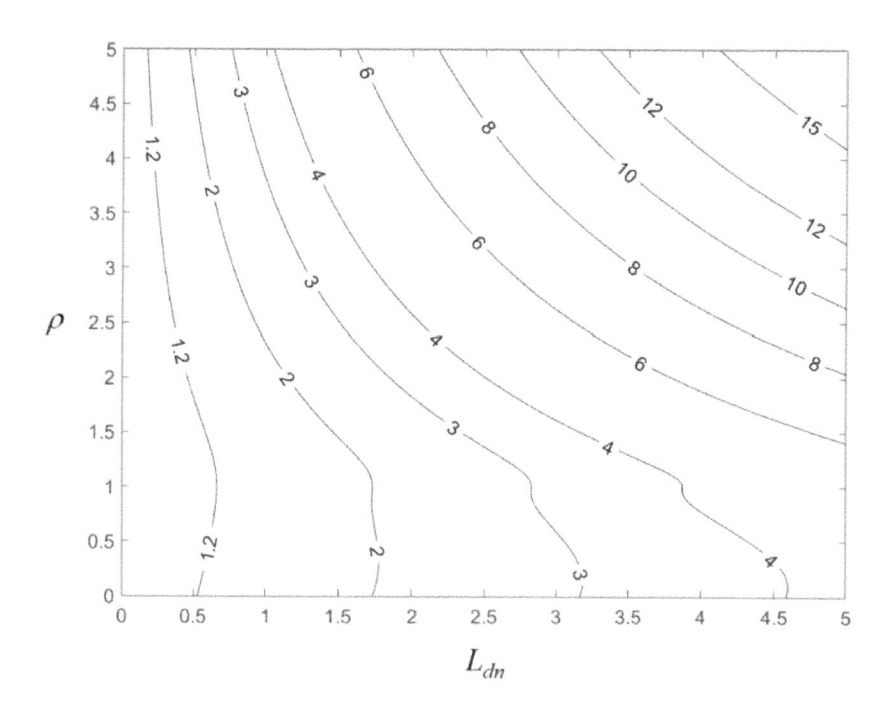

Figure 2.13. Values of $V_{n\,max}$ in the (L_{dn}, ρ) plane for the lossless model

Finally, it should be noted that the normalized excitation current resistance value is constant and equal to 1 by definition.

2.4.1.3. A special case of a variable-reluctance synchronous machine

Variable-reluctance synchronous machines are not excited. Therefore, it is impossible to choose normalized quantities related to the excitation flux. In the following, we will redefine the normalized quantities and parameters by rewriting them as:

$$L_{dn} = \frac{L_d \cdot I_{max} \cdot \omega_b}{V_{max}} \qquad [2.37]$$

$$V_n = \frac{V}{V_{max}} \tag{2.38}$$

$$R_{sn} = \frac{R_s \cdot I_{max}}{V_{max}} \text{ and } R_{fn} = \frac{R_f \cdot I_{max}}{V_{max}} \tag{2.39}$$

where ω_b is the electric angular frequency of the base speed.

Moreover, the absence of an excitation flux does not allow for a distinguished direct axis to be defined. For excited machines, the direct axis is that of the maximal excitation flux. For our developments, we choose to define the direct axis as the one along which the inductance is minimal for the first harmonic model.

Table 2.6 provides some variational ranges of the different normalized quantities and parameters in this new framework.

Quantities and parameters	Variational intervals
I_n	[0, 1]
V_n	[0, 1]
P_n	[0, 1]
Ω_n	[0, +∞)
Γ_n	[0, 1]
ρ	(1, +∞)
L_{dn}	(0, +∞)
R_{sn}	[0, 1]
R_{fn}	(0, +∞)

Table 2.6. *Variational intervals of the normalized quantities and parameters for variable-reluctance synchronous machines*

At low speeds, in order to impose the maximum torque for these machines, the MTPA control law is adopted just like before, while the maximum current I_{max} (nominal current) is imposed. Equation [2.40] provides the conditions for this case, as well as some quantities for the lossless model. As before, it can be shown that it is more interesting to operate in the so-called stable region (see the appendix: https://1drv.ms/b/s!AogSAGtYvycUkHIbc3FbsPoSQjJt?e=U9p85C).

$$\begin{cases} I_n = 1 \\ \Psi_{opt} = \dfrac{\pi}{4} \\ V_n = \Omega_n \cdot I_n \cdot \dfrac{L_{dn}}{\sqrt{2}} \cdot \sqrt{\rho^2 + 1} \\ PF = \left(\dfrac{(\rho - 1)}{\sqrt{2} \cdot \sqrt{\rho^2 + 1}} \right) \end{cases} \qquad [2.40]$$

The normalized inductance for the lossless model is then defined by:

$$L_{dn} = \sqrt{\dfrac{2}{\rho^2 + 1}} \qquad [2.41]$$

Figure 2.14 shows the normalized inductance variation along the direct axis and the power factor for the MTPA operation as a function of the saliency ratio ρ for the lossless model. The power factor tends to the value of 0.7071 when the salience ratio ρ tends to infinity (Betz 1992; Betz et al. 1993). Excited synchronous machines provide better power factors (Figure 2.12(b)).

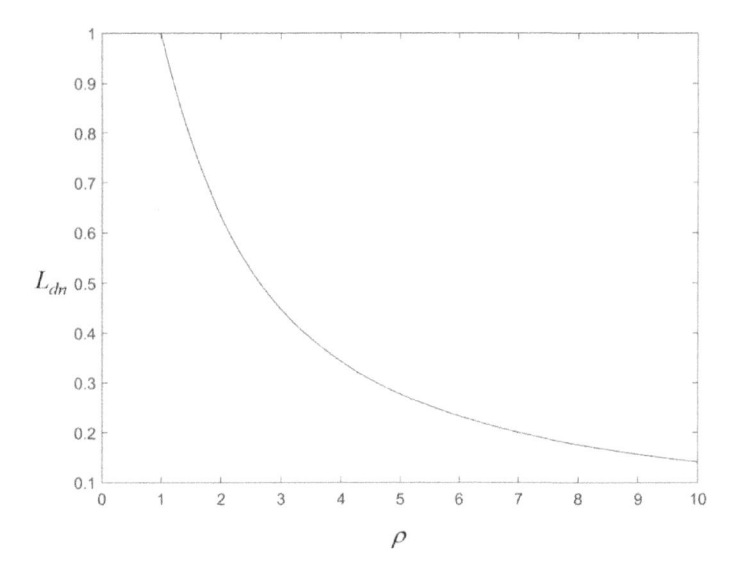

a) Normalized inductance along the direct axis

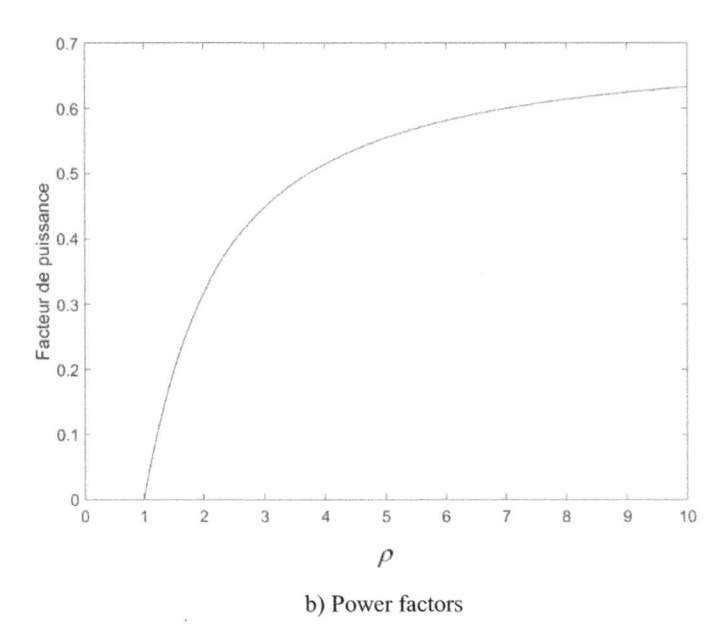

Figure 2.14. *Normalized inductance along the direct axis and the power factors for variable-reluctance synchronous machines (lossless model)*

2.4.2. Operation at high speeds/the notion of flux weakening

At low speeds, the maximum torque is obtained by adopting the MTPA control law and imposing the maximum amplitudes for the armature current and the excitation field flux. Under these conditions, we have seen that the maximum operational speed corresponds to the base speed. At this speed, the armature voltage amplitude attains its maximum value.

To be able to operate beyond this speed while still maximizing the torque, it is necessary to implement control laws that allow for the flux of a synchronous machine to be weakened in an optimal way. Optimality here corresponds to maximizing the torque at these speeds.

For the lossless model, the need to weaken the flux of the machine comes from the fact that the voltage amplitude is proportional to the product of the stator magnetic flux and the angular frequency [2.7]. It is therefore easy to understand the need to reduce the magnetic flux in the stator, in order to operate at high speeds.

Although we have not explicitly shown it, the control law for maximizing the torque corresponds to the MTPV law. This is explicitly demonstrated in the appendix (https://1drv.ms/b/s!AogSAGtYvycUkHIbc3FbsPoSQjJt?e=U9p85C). The latter allows for the torque to be maximized when it is possible to maximize the amplitudes of the armature voltage and the field flux while fixing the optimal internal angle ([2.15] and [2.16]). However, the magnitude of the resulting armature current is particularly high, especially at low speeds.

When it is not possible to fix the optimal angle given by the expressions [2.15] and [2.16] while still respecting the rated current constraints, other control laws must be implemented instead.

2.4.2.1. Circle diagrams

Circle diagrams are useful graphical tools for analyzing how electrical machines operate. In a book titled "Mémoires sur l'électricité et l'optique" by Alfred Potier (1840–1905), published in 1912, prefaced by Henri Poincaré (1854–1912) and annotated by André Blondel (1863–1938), André Blondel attributes the first use of these diagrams to Alfred Potier. For readers interested in the history of science, it should be noted that these memoirs were first published in 1894.

These diagrams are always used for different electrotechnical objects (Terman 1926; Kron 1930; Terman et al. 1930). For our purposes, they represent the operational characteristics of a synchronous machine, both visually and graphically (Amara 2001; De Doncker et al. 2011; Nam 2019).

We will use them in this work to graphically illustrate the trajectories of different control laws and strategies in the (i_{dn}, i_{qn}) plane (Amara 2001; De Doncker et al. 2011; Nam 2019).

2.4.2.1.1. Maximum armature current circle

This circle is defined by:

$$i_{dn}^2 + i_{qn}^2 = 1 \qquad [2.42]$$

2.4.2.1.2. Maximum armature voltage ellipse

This ellipse is defined by:

$$(\rho \cdot L_{dn} \cdot i_{qn})^2 + (k_f + L_{dn} \cdot i_{dn})^2 = \left(\frac{V_{n\ max}}{\Omega_n}\right)^2 \quad [2.43]$$

and the center of this ellipse is given by: $(i_{dn}, i_{qn}) = \left(\frac{-k_f}{L_{dn}},\ 0\right)$.

Note that the center of the ellipse lies outside of the circle delimited by the armature current when $L_{dn} < 1$, on the circumference when $L_{dn} = 1$, and inside the circle when $L_{dn} > 1$. Furthermore, depending on the value of the saliency ratio, we can distinguish three cases:

– $\rho < 1$: the major axis of the ellipse is parallel to the $(i_{dn} = 0)$ axis (i_{qn}-axis);

– $\rho = 1$: the ellipse is a circle;

– $\rho > 1$: the major axis of the ellipse is carried by the line $i_{qn} = 0$ (i_{dn}-axis).

2.4.2.1.3. "Iso-torque" curves

These curves can be deduced directly from the normalized torque expression defined by:

$$i_{qn} = \frac{\Gamma_n \cdot V_{n\ max}}{(k_f + (1-\rho) \cdot L_{dn} \cdot i_{dn})} \quad [2.44]$$

2.4.2.1.4. MTPA trajectory

For a given armature current amplitude and field flux value, the angle ψ corresponding to the MTPA control law, defined previously by equations [2.12] and [2.13], is expressed as a function of the normalized quantities and parameters as follows:

$$\psi_{opt} = \begin{cases} 0 & \rho = 1 \\ \arcsin\left(\dfrac{k_f - \sqrt{k_f^2 + 8 \cdot (1-\rho)^2 \cdot L_{dn}^2 \cdot I_n^2}}{4 \cdot (1-\rho) \cdot L_{dn} \cdot I_n}\right), \Gamma_n(I_n, \psi_{opt}) > 0 & \rho \neq 1 \end{cases} \quad [2.45]$$

The MTPA trajectory is then defined by the following parametric equations:

$$\begin{cases} i_{dn} = -I_n \cdot \sin(\psi_{opt}) \\ i_{qn} = I_n \cdot \cos(\psi_{opt}) \end{cases} \qquad [2.46]$$

For machines with non-salient poles ($\rho = 1$), this trajectory is given by:

$$\begin{cases} i_{dn} = 0 \\ i_{qn} = I_n \end{cases}$$

For these machines ($\rho = 1$) in motor operation, it is a half-line ($I_n > 0$) along the i_{qn}-axis. Negative values of I_n instead correspond to generator operation.

2.4.2.1.5. MTPV trajectory

For a given armature voltage amplitude and field flux value, the internal angle δ corresponding to the MTPV control law, defined previously by equations [2.15] and [2.16], is expressed as a function of the normalized quantities and parameters as follows:

$$\delta_{opt} = \begin{cases} \dfrac{\pi}{2} & \rho = 1 \\ \arccos\left(\dfrac{-k_f + \sqrt{k_f^2 + 8 \cdot \left(\dfrac{1}{\rho} - 1\right)^2 \cdot \left(\dfrac{V_n}{\Omega_n}\right)^2}}{4 \cdot \left(\dfrac{1}{\rho} - 1\right) \cdot (V_n / \Omega_n)} \right), \; \Gamma_n(I_n, \delta_{opt}) > 0 & \rho \neq 1 \end{cases} \qquad [2.47]$$

The MTPV trajectory is then defined by the following parametric equations:

$$\begin{cases} i_{dn} = \dfrac{V_n \cdot \cos(\delta_{opt}) - k_f \cdot \Omega_n}{L_{dn} \cdot \Omega_n} \\ i_{qn} = \dfrac{V_n \cdot \sin(\delta_{opt})}{\rho \cdot L_{dn} \cdot \Omega_n} \end{cases} \qquad [2.48]$$

For machines with non-salient poles ($\rho = 1$), this trajectory is given by:

$$\begin{cases} i_{dn} = \dfrac{-k_f}{L_{dn}} \\ i_{qn} = \dfrac{V_n}{L_{dn} \cdot \Omega_n} \end{cases}$$

For a given value of the field flux (k_f given), and when the speed varies from at rest to infinity $\Omega_n \in [0, +\infty)$, for these machines ($\rho = 1$) in motor operation ($i_{qn} > 0$), it is a half-line along the i_{qn}-axis. Negative values of i_{qn}, and therefore of V_n, correspond to generator operation.

2.4.2.1.6. UPF (Unity Power Factor) trajectory

For a given armature current amplitude and field flux value, the angle ψ corresponding to the UPF control law, and expressed as a function of the normalized quantities and parameters, is given by equation [2.49]. It should be recalled that this angle exists only under certain conditions (Figure 2.8). The MTPA trajectory is then defined by the parametric equations [2.50].

$$\psi_{UPF} = \begin{cases} \arcsin\left(\dfrac{L_{dn} \cdot I_n}{k_f}\right), \; \Gamma_n(I_n, \psi_{UPF}) > 0 & \rho = 1 \\ \arcsin\left(\dfrac{k_f - \sqrt{k_f^2 + 4\cdot(\rho-1)\cdot\rho\cdot L_{dn}^2 \cdot I_n^2}}{2\cdot(1-\rho)\cdot L_{dn} \cdot I_n}\right) \\ \arcsin\left(\dfrac{k_f + \sqrt{k_f^2 + 4\cdot(\rho-1)\cdot\rho\cdot L_{dn}^2 \cdot I_n^2}}{2\cdot(1-\rho)\cdot L_{dn} \cdot I_n}\right), \; \Gamma_n(I_n, \psi_{UPF}) > 0 \;\; \rho \neq 1 \end{cases} \quad [2.49]$$

$$\begin{cases} i_{dn} = -I_n \cdot \sin(\psi_{UPF}) \\ i_{qn} = I_n \cdot \cos(\psi_{UPF}) \end{cases} \quad [2.50]$$

2.4.2.1.7. Synthesis

To illustrate each curve defined in section 2.4.2.1, we have drawn a circle diagram in the case of a synchronous machine with: $\rho = 0.2$, $L_{dn} = 1.2$ and $V_n = k_f = 1$ (Figure 2.15). Note that there are two different trajectories for the same UPF control law. The area of the maximum voltage ellipses decreases as the speed increases. At base speed, the maximum voltage ellipse is the one intersecting both the maximum current circle and the MTPA trajectory.

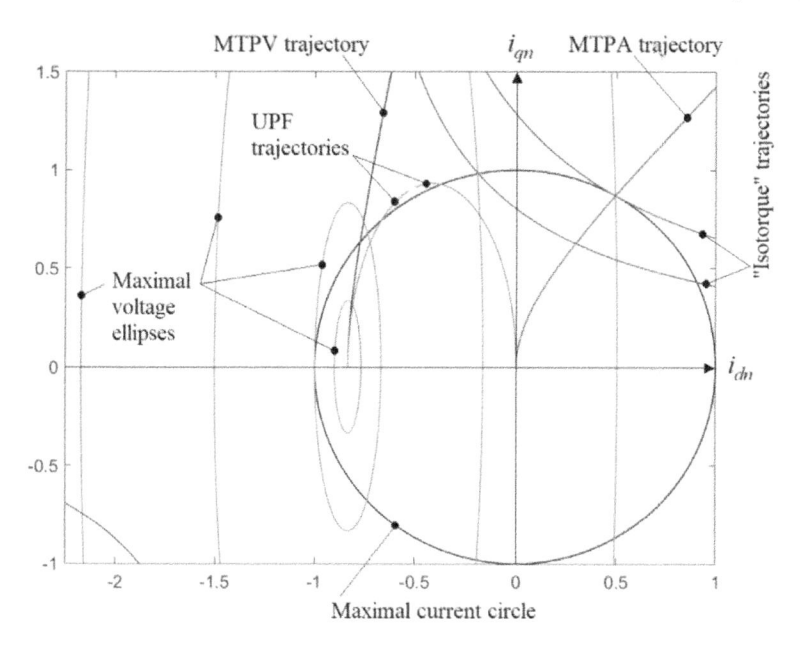

Figure 2.15. *Circle diagram for a synchronous machine with $p = 0.2$, $L_{dn} = 1.2$, for $V_n = k_f = 1$*

2.4.2.2. Machines controls with a constant excitation flux

In order to evaluate the contribution hybrid excitation provides or, more precisely, the possibility of controlling the field flux, we start by studying machines that operate with a constant field flux. This is always the case for permanent magnet machines and can also correspond to machines where wound excitation is present (simple wound excitation machines and hybrid excited machines), such that the excitation current is kept constant. This may be the case in the presence of faults, in which an inability to control the excitation current occurs.

We will repeat the study of the fundamental control laws defined in section 2.3.4, now in the context of high-speed operation ($\Omega_n > 1$), and also define new ones adapted to this operation (Sneyers 1985; Jahns et al. 1986; Jahns 1987). Additionally, we will define the optimal control strategy that maximizes the torque.

As a preamble before studying high-speed operations ($\Omega_n > 1$), in order to maximize the torque for a given speed, it should be remembered that the voltage amplitude must be maintained at its maximal value ($V_n = V_{n\,max}$), as well as the field flux magnitude ($k_f = 1$). On the other hand, for operations at speeds below the base speed ($\Omega_n \leq 1$), it is not possible to maintain the voltage amplitude at its maximal value, without the risk of the armature current amplitude exceeding its limit value. Above the base speed, this amplitude must be maintained at its greatest value.

2.4.2.2.1. Maximum torque per ampere (MTPA) operation

Naturally, it is relevant to consider the maximum torque per ampere (MTPA) control law. For this, it is necessary to fix the angle ψ from equations [2.12] and [2.13], while noting that:

$$V_{n\,max} = \Omega_n \cdot \sqrt{\left(\rho \cdot L_{dn} \cdot I_n \cdot \cos(\psi_{opt})\right)^2 + \left(1 - L_{dn} \cdot I_n \cdot \sin(\psi_{opt})\right)^2}$$

By solving this last equation, we obtain the armature current amplitude variation as a function of speed:

$$I_n = \begin{cases} \dfrac{\sqrt{(V_{n\,max}/\Omega_n)^2 - 1}}{L_{dn}} & \rho = 1 \\[2ex] \sqrt{\dfrac{f(\Omega_n) - 1}{8 \cdot (1-\rho)^2 \cdot L_{dn}^2}} & \rho \neq 1 \end{cases} \quad [2.51]$$

where:

$$f(\Omega_n) = \dfrac{-(3-\rho) \cdot (1-\rho) + 2 \cdot \sqrt{\rho^3 \cdot (4 - 3 \cdot \rho) + (1+\rho^2) \cdot (V_{n\,max}/\Omega_n)^2 \cdot 4 \cdot (1-\rho)^2}}{(1+\rho^2)}$$

Note that, for machines with non-salient poles ($\rho = 1$), we have:

$$L_{dn} = \sqrt{V_{n\,max}^2 - 1}$$

It can be shown that $I_n(\Omega_n)$ is a decreasing function of Ω_n. The current magnitude is then less than the maximum magnitude ($I_n \leq 1$), and the maximum speed that can be achieved by adopting this control law is given by:

$$\Omega_{n\ max\ MTPA} = V_{n\ max} \qquad [2.52]$$

The amplitude of the current cancels out at this speed, and this speed is all the greater still since the values of the torque (L_{dn}, ρ) are large (Figure 2.13).

2.4.2.2.2. Maximum torque per volt (MTPV) operation

This involves fixing the internal angle δ given by equations [2.15] and [2.16], while imposing the maximum amplitude of the armature voltage.

The expression for the normalized armature current amplitude is then given by:

$$I_n = \begin{cases} \dfrac{\sqrt{1+\left(V_{n\ max}/\Omega_n\right)^2}}{L_{dn}} & \rho=1 \\[2ex] \dfrac{1}{L_{dn}} \cdot \sqrt{\left(\dfrac{1-3\cdot\rho}{8\cdot\rho\cdot(1-\rho)}\right) \cdot \sqrt{\rho^2+8\cdot(1-\rho)^2\cdot\left(\dfrac{V_{n\ max}}{\Omega_n}\right)^2} + \dfrac{1}{2}\cdot\left(\dfrac{V_{n\ max}}{\Omega_n}\right)^2 \cdot \left(1+\dfrac{1}{\rho^2}\right)+1+\left(\dfrac{3\cdot\rho-1}{8\cdot(1-\rho)}\right)} & \rho \neq 1 \end{cases} \qquad [2.53]$$

It can be shown that this control law only applies to machines with a normalized armature magnetic reaction along the direct d-axis that is greater than 1 ($L_{dn} > 1$), and for speeds greater than a given speed $\Omega_{\lim n}$ [2.54] (see the appendix: https://1drv.ms/b/s!AogSAGtYvycUkHIbc3FbsPoSQjJt?e=U9p85C),

$$\Omega_{\lim n} = \begin{cases} \dfrac{V_{nn\ max}}{\sqrt{L_{dn}^2-1}} & \rho=1 \\[2ex] \dfrac{V_{n\ max}\cdot|1-\rho|\cdot 2\cdot\sqrt{2}}{\rho\cdot\sqrt{x^2-1}} & \rho \neq 1 \end{cases} \qquad [2.54]$$

where:

$$x = \frac{2 \cdot \sqrt{(4 \cdot \rho - 3) + 4 \cdot (\rho^2 + 1) \cdot (1 - \rho)^2 \cdot L_{dn}^2} - (3 \cdot \rho^2 - 4 \cdot \rho + 1)}{(\rho^2 + 1)}$$

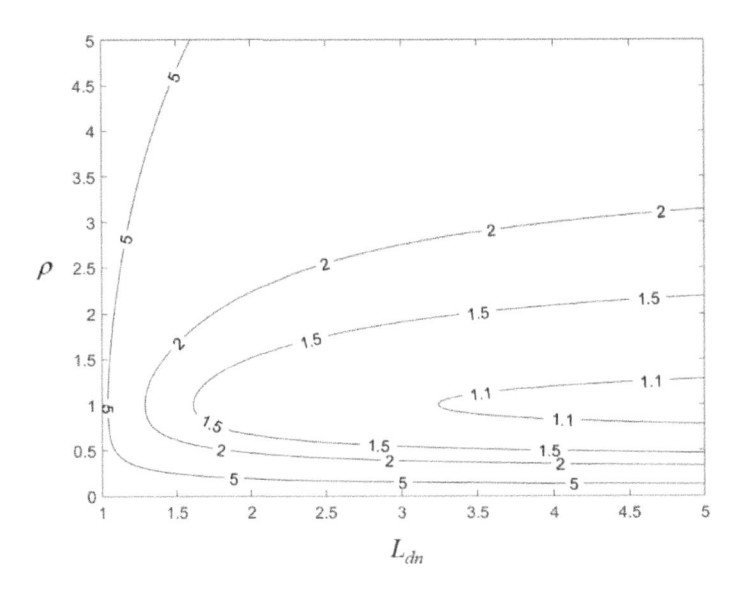

Figure 2.16. Isovalues of $\Omega_{lim\,n}$ in the $(L_{dn} > 1, \rho)$ plane for the lossless model

In this case, the maximum speed is theoretically infinite. The normalized current is strictly decreasing with respect to the speed and tends to the inverse of L_{dn} in this case.

$$\lim_{\Omega_n \to +\infty} I_n = \frac{1}{L_{dn}} \qquad [2.55]$$

Hence, it is clear why this control law cannot be applied to machines with $L_{dn} < 1$. The speed $\Omega_{lim\,n}$ corresponds to that for which $I_n = 1$.

Figure 2.16 shows the speed isovalues for $\Omega_{\lim n}$ in the $(L_{dn} > 1, \rho)$ plane. For a given saliency ratio value ρ, the greater the normalized inductance L_{dn}, the closer this speed is to the base speed.

2.4.2.2.3. Unity power factor (UPF) operation

This involves fixing the angle ψ given by equations [2.25] and [2.26], while noting that:

$$V_{n\ max} = \Omega_n \cdot \sqrt{\left(\rho \cdot L_{dn} \cdot I_n \cdot \cos(\psi_{UPF})\right)^2 + \left(1 - L_{dn} \cdot I_n \cdot \sin(\psi_{UPF})\right)^2}$$

By solving this last equation, we obtain the armature current amplitude variation as a function of speed:

$$I_n = \begin{cases} \dfrac{1}{L_{dn}} \cdot \sqrt{1 - \left(\dfrac{V_{n\ max}}{\Omega_n}\right)^2} & \rho = 1 \\[2ex] \dfrac{1}{L_{dn}} \cdot \sqrt{\dfrac{\left(\rho - 2 \cdot \left(\dfrac{V_{n\ max}}{\Omega_n}\right)^2 + \sqrt{\rho^2 + 4 \cdot (1-\rho) \cdot \left(\dfrac{V_{n\ max}}{\Omega_n}\right)^2}\right)}{2 \cdot \rho}} & \rho \neq 1 \end{cases} \quad [2.56]$$

Figure 2.17 presents the speed ranges for which this control law is applicable according to the parameter values (L_{dn}, ρ). Expressions of the various characteristic speeds indicated in this figure are given by equation [2.57].

$$\begin{cases} \Omega_{UPFn0} = V_{n\ max} \\[1ex] \Omega_{UPFn1} = V_{n\ max} \cdot \sqrt{\dfrac{2}{1 - \rho \cdot 2 \cdot L_{dn}^2 + \sqrt{1 + 4 \cdot \rho \cdot L_{dn}^2 \cdot (\rho - 1)}}} \\[3ex] \Omega_{UPFn2} = V_{n\ max} \cdot \sqrt{\dfrac{2}{1 - \rho \cdot 2 \cdot L_{dn}^2 - \sqrt{1 + 4 \cdot \rho \cdot L_{dn}^2 \cdot (\rho - 1)}}} \\[3ex] \Omega_{UPFn3} = \dfrac{2 \cdot V_{n\ max} \cdot \sqrt{\rho - 1}}{\rho} \end{cases} \quad [2.57]$$

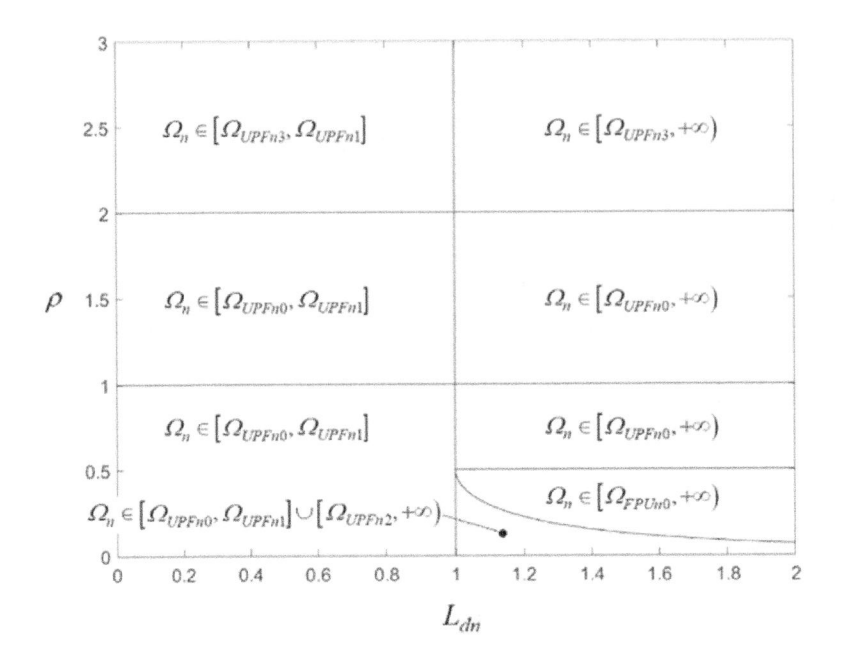

Figure 2.17. *Speed ranges for which the UPF control law is applicable in the (L_{dn}, ρ) plane*

These expressions follow from the existence of solutions for the equation given at the beginning of this section representing the voltage magnitude, in addition to the resulting current magnitude [2.56]. For Ω_{UPFn0}, the armature current amplitude is zero. For the normalized speeds Ω_{UPFn1} and Ω_{UPFn2}, the normalized armature current amplitude is 1 ($I_n = 1$). Finally, the normalized speed Ω_{UPFn3} corresponds to the value for which solutions to the voltage amplitude equation exist.

In Figure 2.17, we may note a clear difference between machines with a normalized inductance of $L_{dn} < 1$, for which the range of unity power factor operating speeds is finite, and those with normalized inductance $L_{dn} \geq 1$, for which the range of unity power factor operating speeds is infinite. This is mainly due to the armature current amplitude variation, which tends to the inverse of L_{dn} when the speed tends to infinity.

For machines with ($L_{dn} > 1$, $\rho < 0.5$), we may distinguish between two zones, whose boundary is given by the equation:

$$L_{dn} = L_{dn0} = \frac{1}{2 \cdot \sqrt{\rho \cdot (1-\rho)}} \qquad [2.58]$$

For machines with ($L_{dn0} > L_{dn} > 1$, $\rho < 0.5$), the speed range for which the control law is applicable is discontinuous. In the first range $\Omega_n \in [\Omega_{UPFn0}, \Omega_{UPFn1}]$, the armature current amplitude is strictly increasing, whereas in the second range $\Omega_n \in [\Omega_{UPFn2}, +\infty]$, it is strictly decreasing.

For machines with ($L_{dn} \geq L_{dn0}$, $\rho < 0.5$), the speed range for which the control law is applicable is again no longer discontinuous, but two operating ranges can be distinguished as before: the first range with $\Omega_n \in [\Omega_{UPFn0}, \Omega_{UPFn4}]$, for which the armature current amplitude is strictly increasing, and the second range with $\Omega_n \in [\Omega_{UPFn4}, +\infty]$, for which the armature current amplitude is strictly decreasing.

The expression for the normalized speed Ω_{UPFn4} is given by:

$$\Omega_{UPFn4} = 2 \cdot V_{n\,max} \cdot \sqrt{\frac{1-\rho}{1-2\cdot\rho}} \qquad [2.59]$$

Note that the angle ψ_{UPF} is equal to $\arcsin(x_1)$ [2.26], for the first operational range, i.e. $\Omega_n \in [\Omega_{UPFn0}, \Omega_{UPFn1}]$ for ($L_{dn0} > L_{dn} > 1$, $\rho < 0.5$), or $\Omega_n \in [\Omega_{UPFn0}, \Omega_{UPFn4}]$ for ($L_{dn} \geq L_{dn0}$, $\rho < 0.5$), and to $\arcsin(x_2)$ [2.26], for the second range, i.e. $\Omega_n \in [\Omega_{UPFn2}, +\infty)$ for ($L_{dn0} > L_{dn} > 1$, $\rho < 0.5$), or $\Omega_n \in [\Omega_{UPFn4}, +\infty)$ for ($L_{dn} \geq L_{dn0}$, $\rho < 0.5$).

Figure 2.18 shows the isovalues of the normalized characteristic velocities Ω_{UPFn0} ($\rho < 2$) and Ω_{UPFn03} ($\rho \geq 2$), in the (L_{dn}, ρ) plane. Figure 2.19 shows the isovalues of the normalized characteristic velocities Ω_{UPFn1}, for ($L_{dn} < 1$) (Figure 2.19(a)) and ($1 \leq L_{dn} < L_{dn0}$, $\rho < 0.5$) (Figure 2.19(b)), and Ω_{UPFn04}, for ($L_{dn} > L_{dn0}$, $\rho < 0.5$) (Figure 2.19(b)). Finally, Figure 2.20 shows the isovalues of the normalized characteristic velocities Ω_{UPFn2} ($1 \leq L_{dn} < L_{dn0}$, $\rho < 0.5$) and Ω_{UPFn04} ($L_{dn} > L_{dn0}$, $\rho < 0.5$). The boundaries between the parameter sets (L_{dn}, ρ) of machines with different behaviors are drawn in these figures.

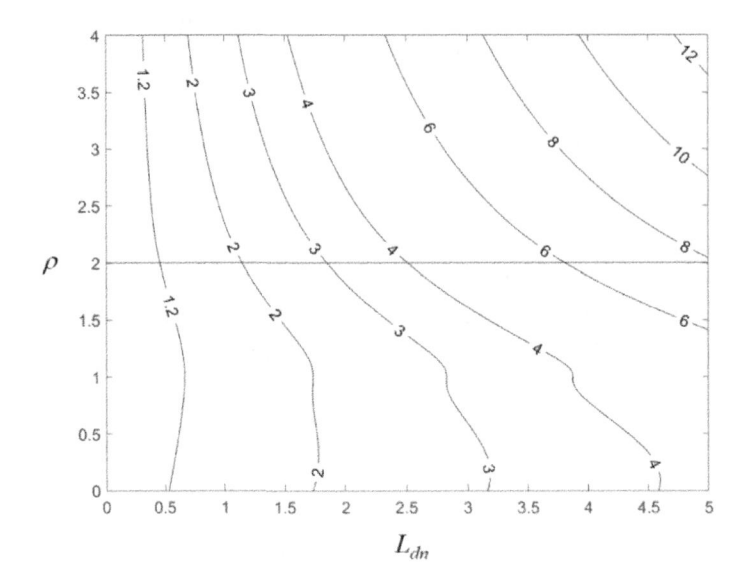

Figure 2.18. *Values of Ω_{UPFn0} ($\rho < 2$) and Ω_{UPFn3} ($\rho \geq 2$) in the (L_{dn}, ρ) plane*

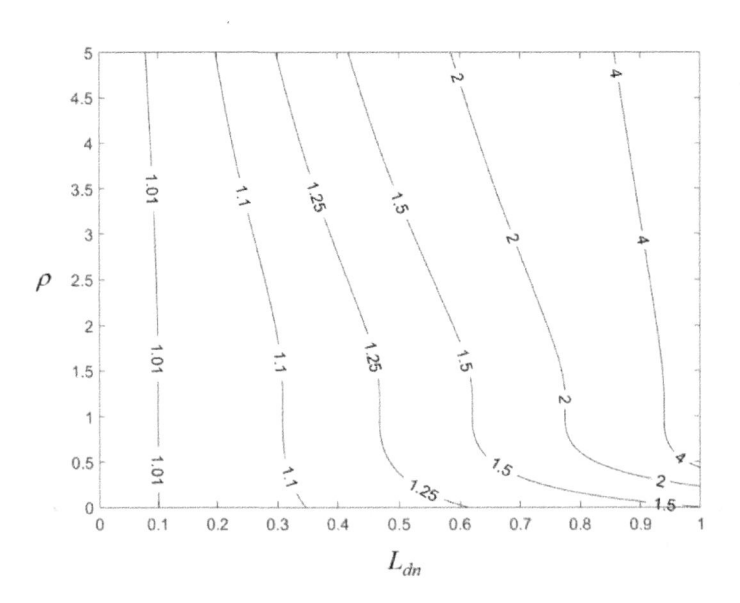

a) Values of Ω_{UPFn1} ($L_{dn} < 1$)

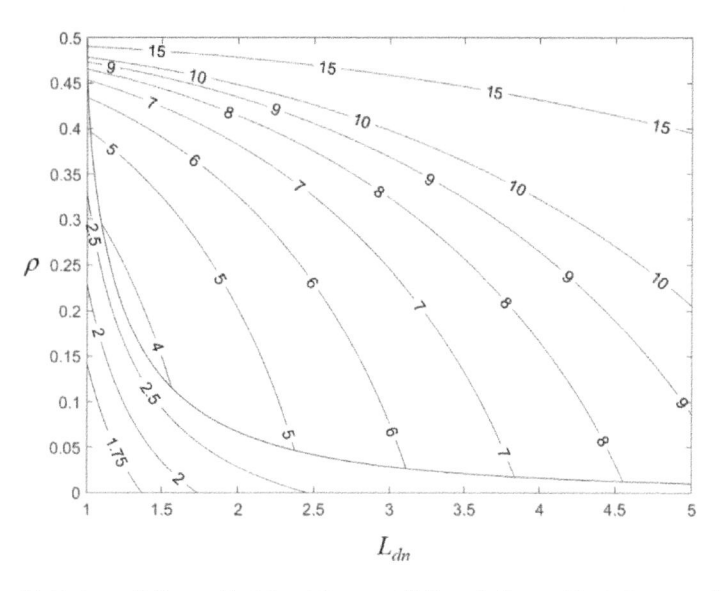

b) Values of Ω_{UPFn1} ($1 \leq L_{dn} < L_{dn0}$, $\rho < 0.5$) and Ω_{UPFn4} ($L_{dn} > L_{dn0}$, $\rho < 0.5$)

Figure 2.19. Values of Ω_{UPFn1} and Ω_{UPFn4} in the (L_{dn}, ρ) plane

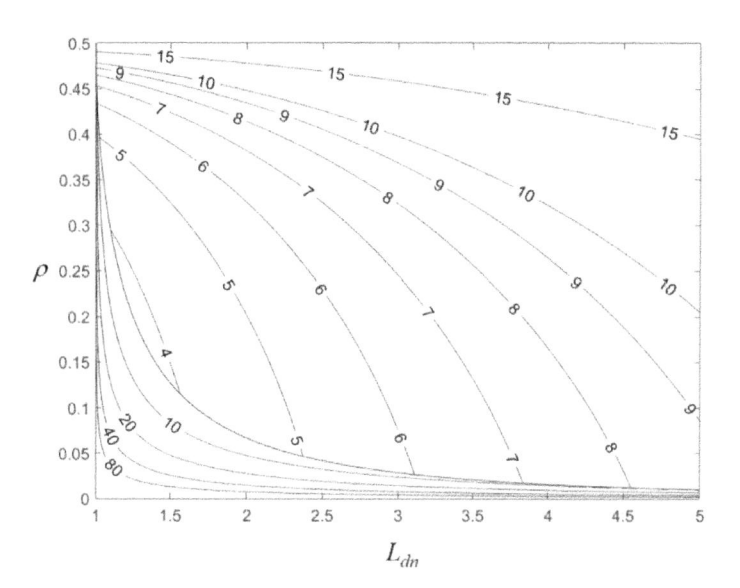

Figure 2.20. Values of Ω_{UPFn2} and Ω_{UPFn4} in the (L_{dn}, ρ) plane

2.4.2.2.4. Maximum current (MC) operation (Kron 1930; Terman et al. 1930; Sneyers 1985)

This control law aims to maintain the amplitude of the armature current at its maximum value $I_n = 1$ (Kron 1930; Terman et al. 1930; Sneyers 1985). To determine the angle ψ to be imposed, we must solve the following equation:

$$V_{n\,\max} = \Omega_n \cdot \sqrt{\left(\rho \cdot L_{dn} \cdot \cos(\psi)\right)^2 + \left(1 - L_{dn} \cdot \sin(\psi)\right)^2}$$

The solution to this equation is the angle ψ given by:

$$\psi_{MC} = \begin{cases} \arcsin\left(\dfrac{L_{dn}^2 + 1 - \left(V_{n\,\max}^2/\Omega_n^2\right)}{2 \cdot L_{dn}}\right) & \rho = 1 \\[2ex] \arcsin\left(\dfrac{1 - \sqrt{\rho^2 \cdot \left(1 + (\rho^2 - 1) \cdot L_{dn}^2\right) + \left(1 - \rho^2\right) \cdot \left(V_{n\,\max}/\Omega_n\right)^2}}{L_{dn} \cdot \left(1 - \rho^2\right)}\right) & \rho \neq 1 \end{cases} \quad [2.60]$$

The maximum speed that can be achieved by adopting this control law depends on the parameter set (L_{dn}, ρ). Table 2.7 summarizes the different cases along with their maximum achievable speeds.

	$\rho < 1$	$\rho \geq 1$
$L_{dn} \leq L_{dn1} \Rightarrow \Omega_{nMC\,\max} = \dfrac{V_{n\,\max}}{\|L_{dn} - 1\|}$		$\Omega_{nMC\,\max} = \dfrac{V_{n\,\max}}{\|L_{dn} - 1\|}$
$L_{dn} > L_{dn1} \Rightarrow \Omega_{nMC\,\max} = \left(\dfrac{V_{n\,\max}}{\rho}\right) \cdot \sqrt{\dfrac{(\rho^2 - 1)}{1 - (1 - \rho^2) \cdot L_{dn}^2}}$		

Table 2.7. *Maximum speeds for the MC control law*

The expression for the normalized inductance L_{dn1} is given by:

$$L_{dn1} = \frac{1}{(1 - \rho^2)} \qquad [2.61]$$

Figures 2.21(a) and (b) show the variations of L_{dn1} as a function of the saliency ratio ρ, and the isovalues of the speed Ω_{nMCmax} in the (L_{dn}, ρ) plane, respectively. Note that $L_{dn1} > 1$, and its value tends to unity when ρ tends to zero.

For machines with ($L_{dn} > L_{dn1}$, $\rho < 1$), the maximum speed corresponds to a speed beyond which it is not possible to respect the limits of both the current ($I_n \leq 1$) and the voltage ($V_n \leq V_{n\ max}$) simultaneously. The previous equation has no solution. For the other machines, the maximum speed corresponds to the speed for which the power tends to be canceled out.

It is important to note that the speed Ω_{nMCmax} is greater than $\Omega_{\lim\ n}$ for $L_{dn} > 1$. This is of particular importance when we define the optimal control strategy, which is the subject of the next section. Moreover, note that the closer L_{dn} is to 1, the larger the value of the normalized speed Ω_{nMCmax}.

2.4.2.2.5. Optimal control strategy

At low speeds ($\Omega \leq \Omega_b \Leftrightarrow \Omega_n \leq 1$), the control law that maximizes torque while respecting current and voltage limits is the MTPA law.

At high speeds ($\Omega > \Omega_b \Leftrightarrow \Omega_n > 1$), the optimal control strategy that should be adopted to maximize the torque while respecting the constraints ($V \leq V_{max}$ and $I \leq I_{max}$) depends on the value of the normalized inductance along the direct d-axis, L_{dn}.

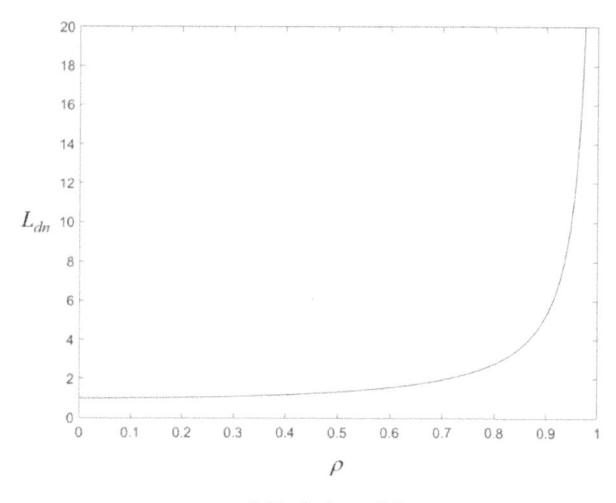

a) Variation of L_{dn1}

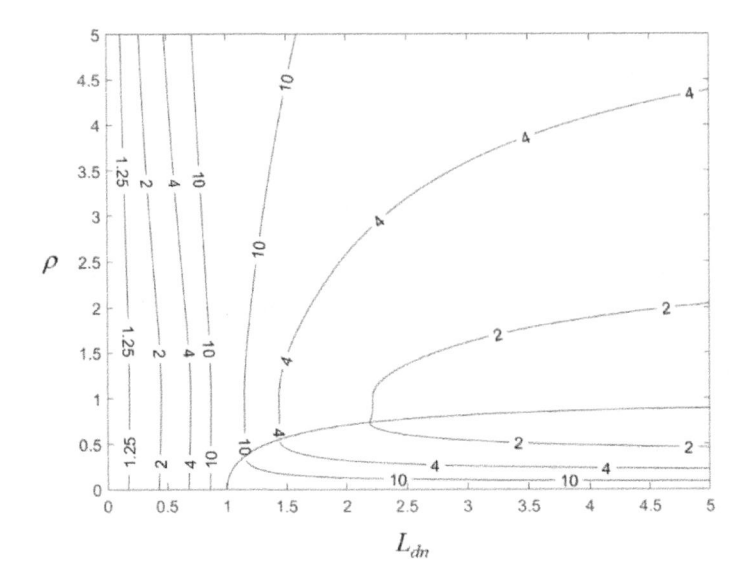

b) Values of Ω_{nMCmax} in the $(L_{dn} \leq L_{dn1}, \rho)$ and $(L_{dn} > L_{dn1}, \rho < 1)$ planes

Figure 2.21. *Variation of L_{dn1} and isovalues of Ω_{nMCmax} in the (L_{dn}, ρ) plane*

We will now distinguish three cases, depending on whether $L_{dn} < 1$, $L_{dn} = 1$ or $L_{dn} > 1$.

2.4.2.2.5.1. $L_{dn} < 1$

For these machines, whether they have non-salient or salient poles, the optimal control strategy is to adopt the MC control law. The maximum speed that can be achieved is given in Table 2.7. The angle ψ at this speed is equal to $\pi/2$.

For these machines with $L_{dn} \leq 1$, the MTPV control law is not applicable as it does not meet the current constraint. It can be shown that the torque produced by adopting the MC control law is greater than the torque produced if either the MTPA or UPF control laws were adopted (see the appendix: https://1drv.ms/b/s!AogSAGtYvycUkHIbc3FbsPoSQjJt?e=U9p85C).

Figure 2.22(a) shows the variation of the maximum normalized power as a function of the normalized speed, for a machine with $L_{dn} = 0.8$ and $\rho = 1.5$. The normalized speed Ω_{PCn} corresponds to the speed $(\Omega_n > 1)$ at which the power is equal to that of the base speed ($\Omega_n = 1$). The operating range over which it would be possible to operate at a constant power is $\Omega_n \in [1, \Omega_{PCn}]$ in this case. The value of Ω_{PCn} can be obtained by solving the following equation:

$$A \cdot \Omega_n^4 + B \cdot \Omega_n^3 + C \cdot \Omega_n^2 + D \cdot \Omega_n + E = 0 \qquad [2.62]$$

where

$$\begin{cases} A = \dfrac{L_{dn}^2 \cdot (1+\rho)^2 \cdot (L_{dn}^2 - 1)}{V_{n\,max}^4 \cdot \rho^2} \\[2mm] B = \dfrac{2 \cdot (L_{dn}^2 \cdot (1+\rho) - 1) \cdot (\rho+1)^2 \cdot L_{dn} \cdot \sqrt{1 - P_n^2 (\Omega_n = 1)}}{V_{n\,max}^3 \cdot \rho^3} \\[2mm] C = \dfrac{(\rho+1)^2 \cdot \left[(1 - P_n^2 (\Omega_n = 1)) \cdot (\rho+1)^2 \cdot L_{dn}^2 + 2 \cdot L_{dn}^2 \cdot \rho - 1\right]}{V_{n\,max}^2 \cdot \rho^4} \\[2mm] D = \dfrac{2 \cdot (\rho+1)^3 \cdot L_{dn} \cdot \sqrt{1 - P_n^2 (\Omega_n = 1)}}{V_{n\,max} \cdot \rho^4} \\[2mm] E = \dfrac{(\rho+1)^2}{\rho^4} \end{cases}$$

Figure 2.22(b) illustrates how the MC control law is applied for this machine in the (i_{dn}, i_{qn}) plane.

For machines with non-salient poles ($\rho = 1$), the expression for the normalized speed Ω_{PCn} is given by:

$$\Omega_{PCn} = \dfrac{V_{n\,max}^2}{1 - L_{dn}^2} \qquad [2.63]$$

Figure 2.23 shows the isovalues for the normalized speed Ω_{PCn} in the ($L_{dn} < 1$, ρ) plane. Note that the closer the value of the normalized inductance L_{dn} is to 1, the greater the normalized speed Ω_{PCn}.

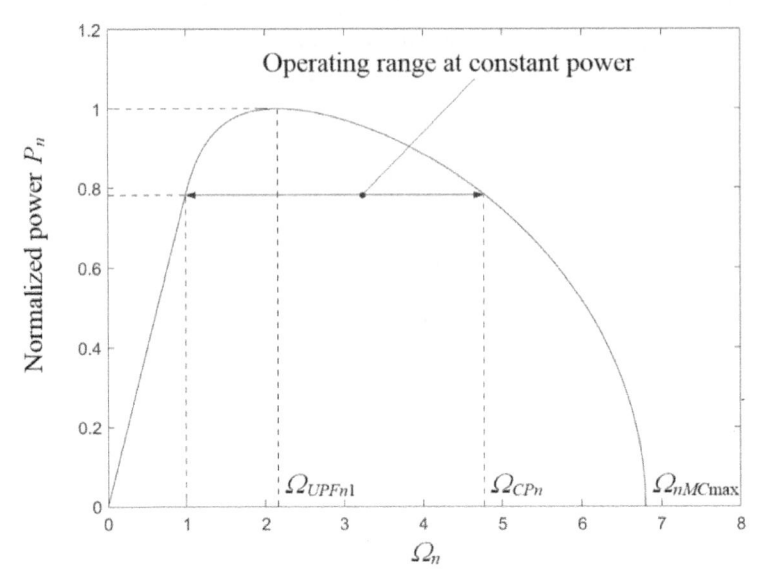

a) Variation of the normalized power as a function of speed ($L_{dn} = 0.8$, $\rho = 1.5$)

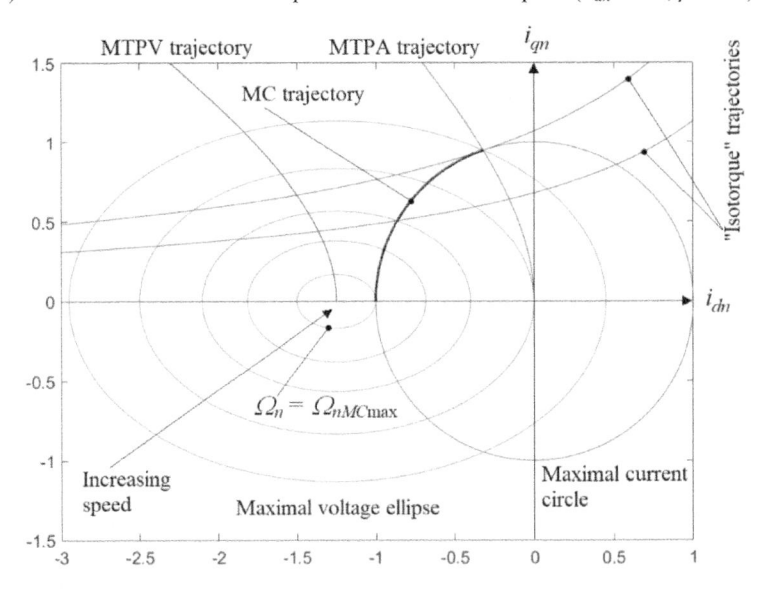

b) Representing the control strategy in the (i_{dn}, i_{qn}) plane.

Figure 2.22. *Variation of the maximum normalized power with speed ($L_{dn} = 0.8$, $\rho = 1.5$) (a), and a representation of the control strategy in the (i_{dn}, i_{qn}) plane (b)*

For a machine with $L_{dn} = 0.8$ and $\rho = 1.5$ (Figure 2.22(a)), we can see that, beyond the base speed, the power increases initially before attaining its maximum, which corresponds to the unity power factor at the normalized speed Ω_{UPFn1}, before then decreasing and becoming zero when the normalized speed Ω_{nMCmax} is reached.

This behavior is shared by a majority of machines with $(L_{dn} < 1, \rho)$. However, some machines with specific parameter sets $(L_{dn} < 1, \rho)$ exhibit different behaviors. Figure 2.24 shows three areas in the $(L_{dn} < 1, \rho)$ plane, allowing us to differentiate between the different behaviors. Indeed, some machines show optimums beyond Ω_{UPFn1}.

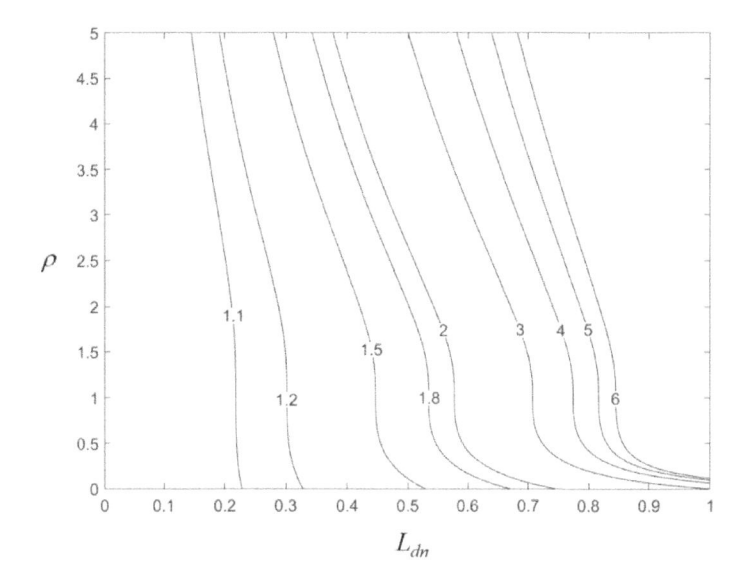

Figure 2.23. *Isovalues for the normalized speed Ω_{PCn}*

The boundary separating "Zone 1" from the two zones "Zone 21" and "Zone 22" is defined by the following relationship:

$$L_{dn} = \sqrt{\frac{4-5\cdot\rho}{4\cdot(1-\rho)\cdot(1-\rho^2)}} \qquad [2.64]$$

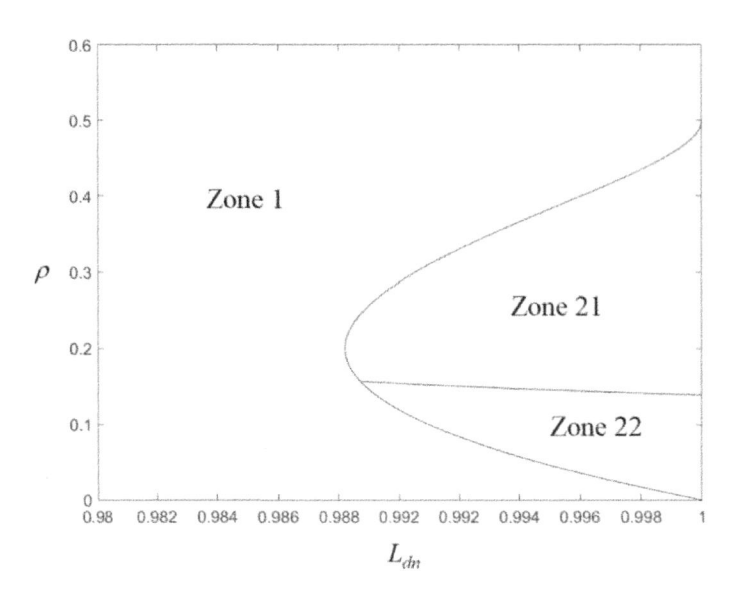

Figure 2.24. Sectorization of the plane ($L_{dn} < 1$, ρ)

Though machines parametrized by (L_{dn}, ρ) belong to "Zone 1", the power strictly decreases beyond the speed Ω_{UPFn1}. For machines with parameters that lie on the boundary, their power has an inflection point that occurs at the normalized speed, given by:

$$\Omega_{Opt0nMC} = \frac{2 \cdot V_{n\_max}}{\rho} \cdot \sqrt{\frac{1-\rho^2}{1-4\cdot\left(1-\left(1-\rho^2\right)\cdot L_{dn}^2\right)}} \qquad [2.65]$$

For other machines in "Zone 21" and "Zone 22", beyond the speed Ω_{UPFn1}, the power initially decreases to a certain speed which we will call $\Omega_{Opt1nMC}$, before then increasing up to a speed which we will call $\Omega_{Opt2nMC}$. Finally, the power decreases until it vanishes at the speed Ω_{nMCmax}. Figure 2.25 shows the normalized power variation for a machine with parameters (L_{dn}, ρ), belonging to "Zone 21".

"Zone 21" corresponds to machines with speeds $\Omega_{PCn} \in (\Omega_{UPFn1}, \Omega_{Opt1nMC})$, and "Zone 22" corresponds to machines with speeds $\Omega_{PCn} \in (\Omega_{Opt2nMC}, \Omega_{nMCmax})$.

The speeds $\Omega_{Opt1nMC}$ and $\Omega_{Opt2nMC}$ can be expressed by:

$$\begin{cases} \Omega_{Opt1nMC} = V_{n\,max} \cdot \sqrt{\dfrac{1-\rho^2}{y_1^2 - \rho^2 \cdot \left(1-\left(1-\rho^2\right)\cdot L_{dn}^2\right)}} \\ \Omega_{Opt2nMC} = V_{n\,max} \cdot \sqrt{\dfrac{1-\rho^2}{y_2^2 - \rho^2 \cdot \left(1-\left(1-\rho^2\right)\cdot L_{dn}^2\right)}} \end{cases} \quad [2.66]$$

where

$$\begin{cases} y_1 = \dfrac{\rho + \sqrt{\rho^2 - 4\cdot\rho\cdot(1-\rho)\cdot\left(1-\left(1-\rho^2\right)\cdot L_{dn}^2\right)}}{2} \\ y_2 = \dfrac{\rho - \sqrt{\rho^2 - 4\cdot\rho\cdot(1-\rho)\cdot\left(1-\left(1-\rho^2\right)\cdot L_{dn}^2\right)}}{2} \end{cases}$$

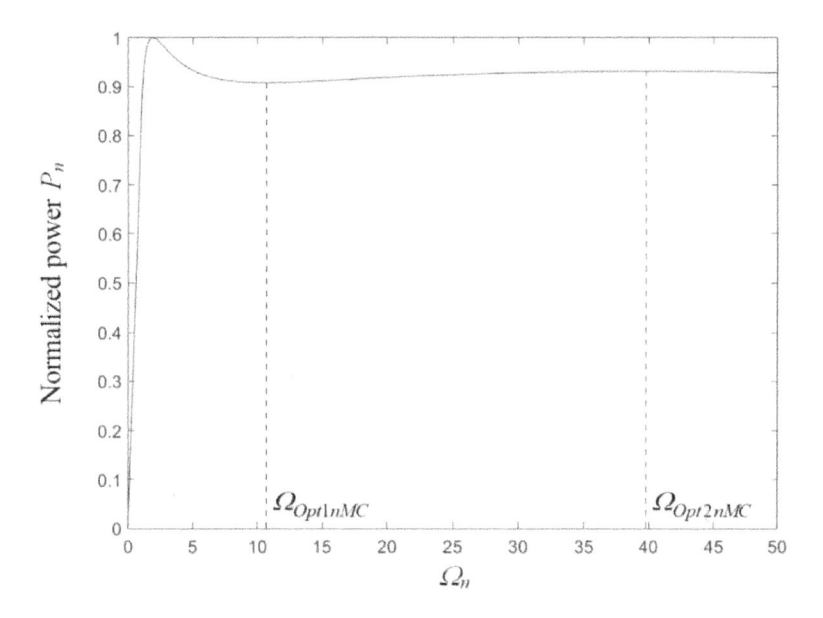

Figure 2.25. *Normalized power variation for (L_{dn} = 0.994, ρ = 0.2) ∈ "Zone 21"*

Figure 2.26 shows the isovalues of the speeds $\Omega_{Opt1nMC}$ and $\Omega_{Opt2nMC}$.

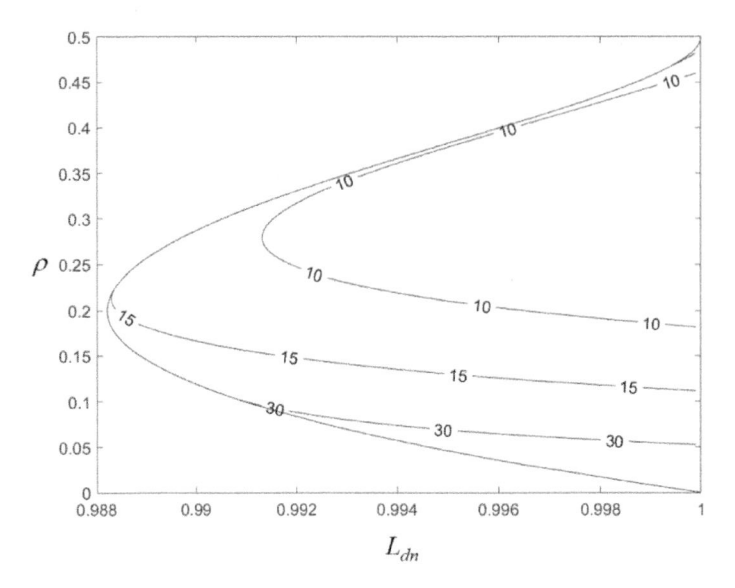

a) Values of $\Omega_{Opt1nMC}$

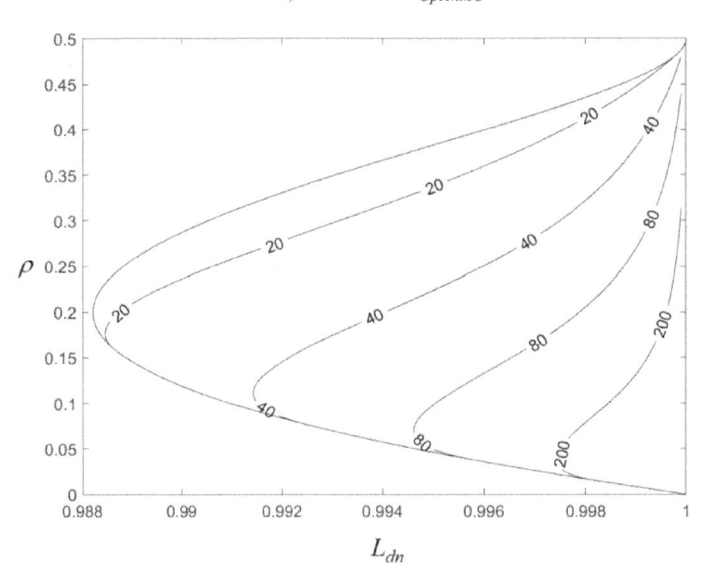

b) Values of $\Omega_{Opt2nMC}$

Figure 2.26. *Values of $\Omega_{Opt1nMC}$ and $\Omega_{Opt2nMC}$ in the (L_{dn}, ρ) plane*

2.4.2.2.5.2. $L_{dn} = 1$

For these machines, the optimal control strategy is to adopt the MC control law. The MPTV control law is not applicable here, because it does not allow the current constraint to be respected. It can be shown that the torque produced by adopting the MC control law is greater than that produced if the MTPA or UPF control laws were adopted instead (see the appendix: https://1drv.ms/b/s!AogSAGtYvycUkHIbc3FbsPoSQjJt?e=U9p85C). Depending on the saliency ratio value, two different behaviors of the function $P_n(\Omega_n)$ can be distinguished.

For machines with $\rho < 0.5$, beyond the base speed, the power initially increases before attaining its maximum ($P_n = 1$) at the speed Ω_{UPFn1}. At this speed, the power factor is unitary. Beyond the speed Ω_{UPFn1}, the power decreases first until the speed reaches $\Omega_{Opt1nMC}$ [2.66], and then increases with the limit:

$$\lim_{\Omega_n \to +\infty} P_n = 1 \qquad [2.67]$$

The maximum speed that can be reached is theoretically infinite, and the same is true with the speed Ω_{PCn} for machines with $\rho > 0.13854$. Figure 2.27 shows how the speed Ω_{PCn} varies as a function of the saliency ratio ρ.

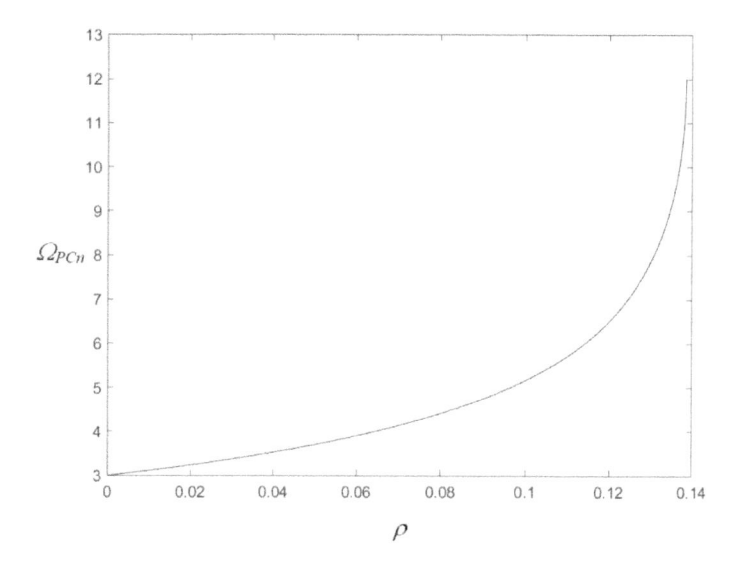

Figure 2.27. Values of Ω_{PCn} as a function of ρ

For these machines, the speed Ω_{UPFn1} can be expressed by:

$$\Omega_{UPFn1} = \frac{V_{n\max}}{\sqrt{1-2\cdot\rho}} \qquad [2.68]$$

For machines with $\rho \geq 0.5$, the normalized power increases and tends towards unity [2.67]. Here again, the maximum speed that can be reached is theoretically infinite, and the same is true for the normalized speed Ω_{PCn}.

2.4.2.2.5.3. $L_{dn} > 1$

For these machines, regardless of whether they have non-salient or salient poles, the optimal control strategy consists of adopting the MC control law up to the speed $\Omega_{\lim n}$, then adopting the MTPV control law beyond that. The maximum speed that can be reached in this case is theoretically infinite. For speeds $\Omega_n \leq \Omega_{\lim n}$, it can be shown that the MC control law can produce greater torque than both the MTPA and UPF control laws (see the appendix: https://1drv.ms/b/s!AogSAGtYvycUkHIbc3FbsPoSQjJt?e=U9p85C).

Depending on the pair of values (L_{dn}, ρ), we can distinguish between three different behaviors of the function $P_n(\Omega_n)$, and Figure 2.28 shows these areas in the (L_{dn}, ρ) plane.

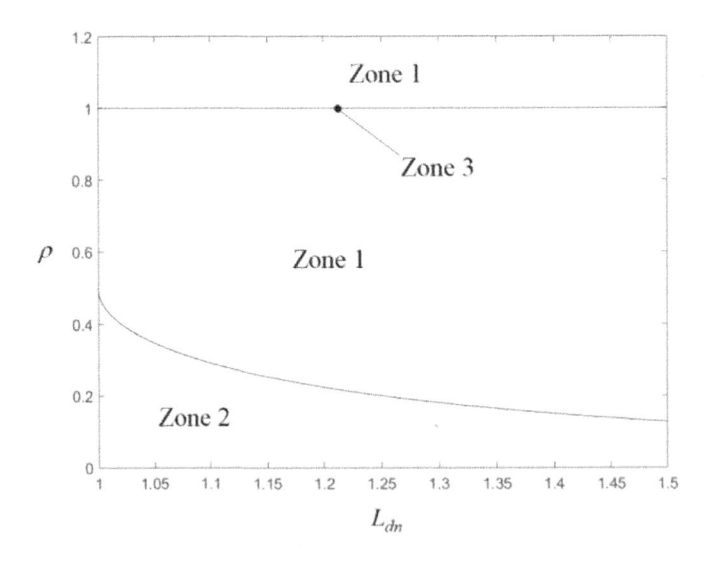

Figure 2.28. *Sectorization of the plane ($L_{dn} > 1$, ρ)*

The boundary between the two zones "Zone 1" and "Zone 2", which was defined in equation [2.58], can also be defined by:

$$\rho = \rho_0 = \frac{L_{dn} - \sqrt{L_{dn}^2 - 1}}{2 \cdot L_{dn}} \qquad [2.69]$$

"Zone 3" corresponds to $\rho = 1$.

For machines belonging to "Zone 1", the power initially increases past the base speed and reaches its maximum at the speed equal to that of $\Omega_{Opt1nMC}$, defined previously in [2.66]. The power then decreases, with the limit:

$$\lim_{\Omega_n \to +\infty} P_n = 1/L_{dn} \qquad [2.70]$$

For machines whose parameters (L_{dn}, ρ) belong to the boundary between the zones "Zone 1" and "Zone 2", the power has the same behavior with a maximum normalized power equal to 1, corresponding to a unitary power factor. The speed for this optimum corresponds to $\Omega_{Opt1nMC}$, and we can show that $\Omega_{Opt1nMC} = \Omega_{UPFn1} = \Omega_{UPFn2}$.

For the machines belonging to "Zone 2", the power initially increases to reach an optimum at the speed Ω_{UPFn1} ($P_n = 1$), before then decreasing to reach a second optimum at the speed $\Omega_{Opt1nMC}$, increasing again to reach a third and final optimum at Ω_{UPFn1} ($P_n = 1$), before finally decreasing to the same limit given by [2.70].

Figure 2.29 shows the isovalues of the speed $\Omega_{Opt1nMC}$ in the plane ($L_{dn} > 1$, ρ), where the boundary between the zones "Zone 1" and "Zone 2" is also represented. While the optimum corresponding to the speed $\Omega_{Opt1nMC}$ is a maximum for machines belonging to "Zone 1", it corresponds to a local minimum for machines belonging to "Zone 2". For machines belonging to "Zone 3" ($\rho = 1$), the power initially increases and reaches its maximum at the speed corresponding to $\Omega_{Opt1nMC} = \Omega_{\lim n}$. Note that for these machines:

$$\Omega_{Opt1nMC} = \Omega_{\lim n} = \frac{V_{n\ max}}{\sqrt{L_{dn}^2 - 1}} \qquad [2.71]$$

Above this speed [2.71], the MTPV control law applies, and the normalized power is constant and equal to:

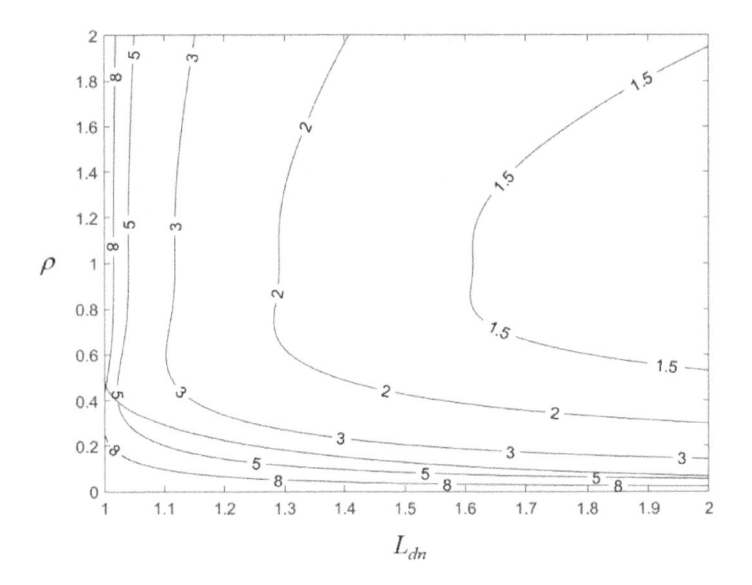

$$P_n(\Omega_n \geq \Omega_{\lim n}) = 1/L_{dn} \qquad [2.72]$$

Figure 2.29. Values of $\Omega_{Opt1nMC}$ in the ($L_{dn} > 1$, ρ) plane

Figure 2.30(a) shows the maximum normalized power variation as a function of normalized speed for a machine with $L_{dn} = 2.5$ and $\rho = 1.5$. Figure 2.30(b) shows the optimal control strategy applied to this machine in the (i_{dn}, i_{qn}) plane for this case. The transition from the MC control law to the MTPV control law is clearly visible.

Figure 2.31 shows another sectorization of the ($L_{dn} > 1$, ρ) plane, based on the possibility of having an infinite ($\Omega_{PCn} = +\infty$, "Zone 1") or a finite ($\Omega_{PCn} < +\infty$, "Zone 2") operational range of constant power.

For machines with $\rho < \rho_0$ [2.69] and whose speed Ω_{PCn} is finite, there are two possibilities:

1) $\Omega_{PCn} \in (\Omega_{UPFn1}, \Omega_{Opt1nMC}]$;

2) $\Omega_{PCn} > \Omega_{UPFn2}$.

Figure 2.32 shows the isovalues for the speed Ω_{PCn}. Note that it is sufficient to have a unit d-axis armature magnetic reaction ($L_{dn} = 1$), for there to be an infinite operational range of constant power.

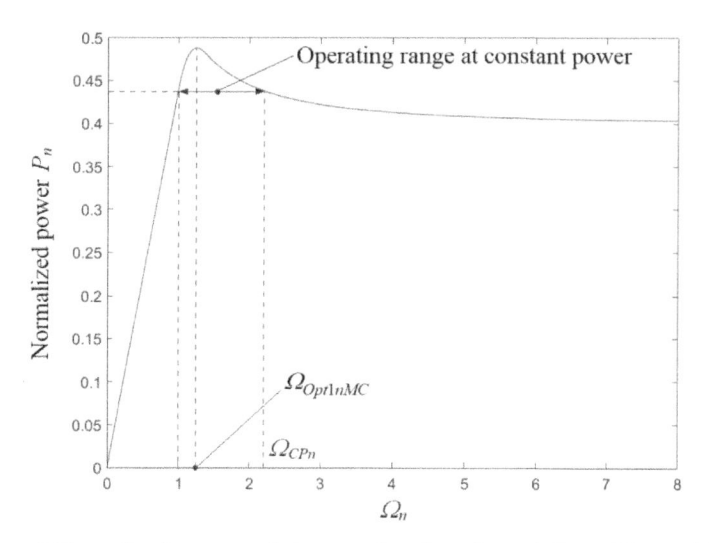

a) Normalized power variation as a function of speed ($L_{dn} = 2.5$, $\rho = 1.5$)

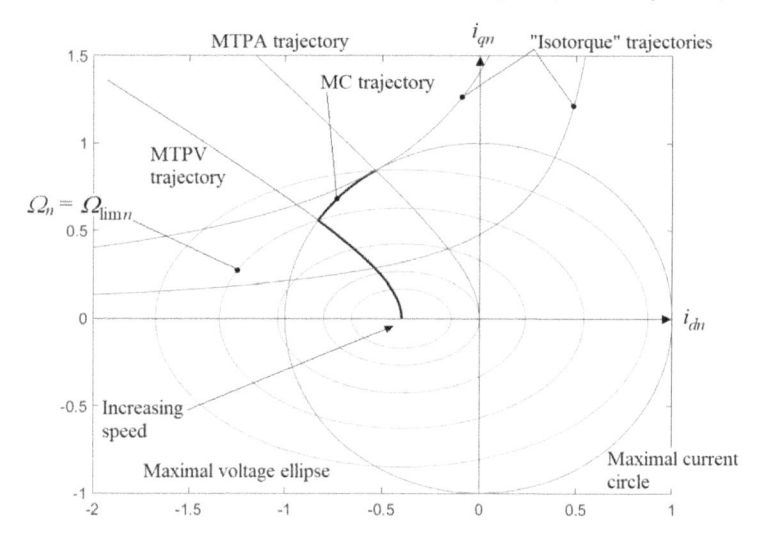

b) Representation of the control strategy in the (i_{dn}, i_{qn}) plane

Figure 2.30. *Maximum normalized power variation of speed ($L_{dn} = 2.5$, $\rho = 1.5$) (a), and a representation of the control strategy in the (i_{dn}, i_{qn}) plane (b)*

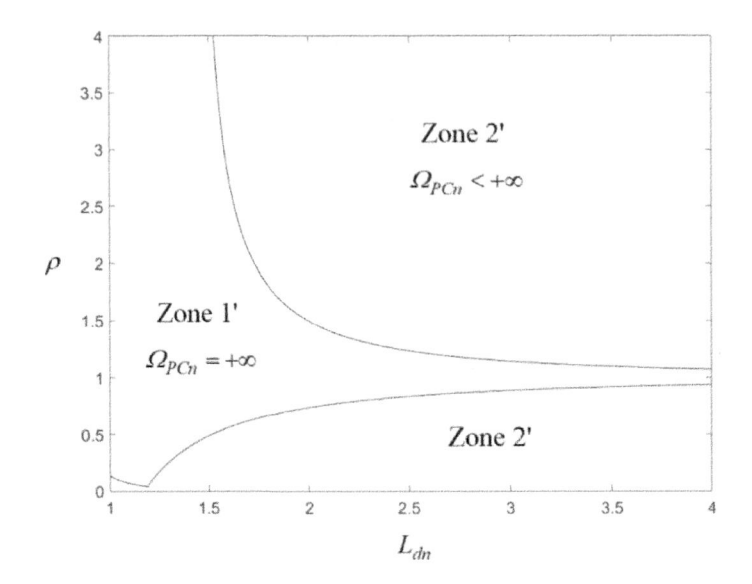

a) Sectorization into zones ($L_{dn} > 1$, ρ)

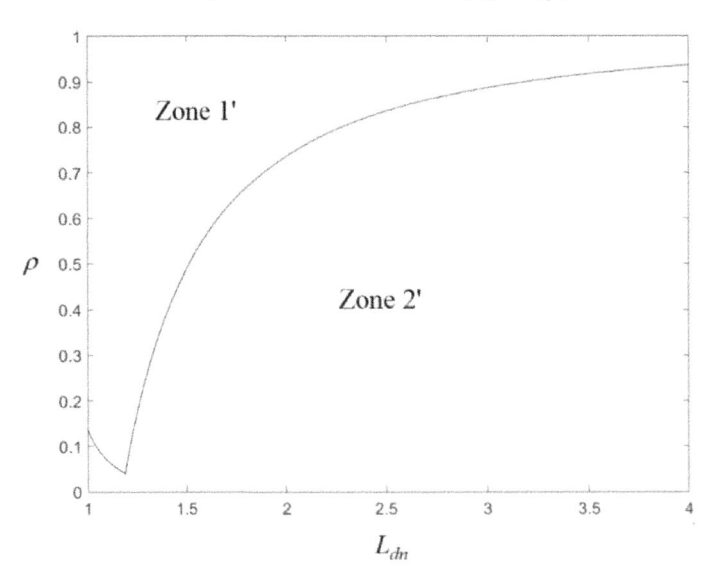

b) Blow-up of the region for small salience ratio values ρ

Figure 2.31. Sectorization of the plane ($L_{dn} > 1$, ρ) depending on whether $\Omega_{PCn} = +\infty$ or $\Omega_{PCn} < +\infty$

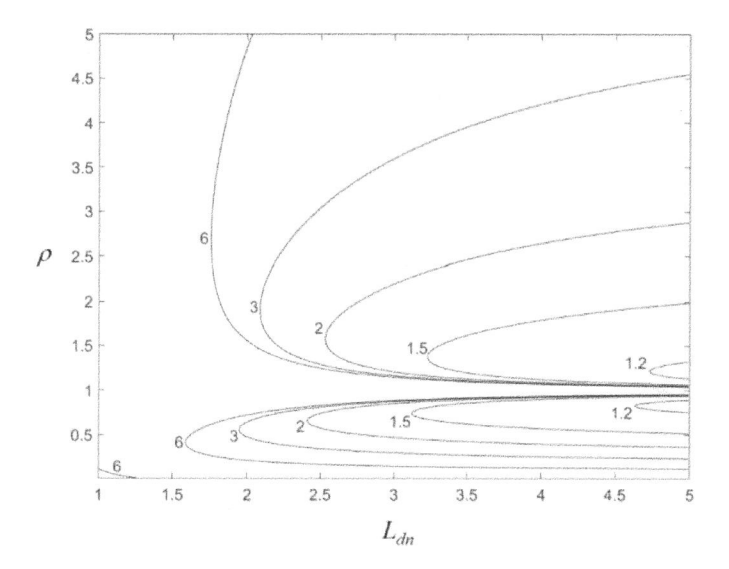

Figure 2.32. *Isovalues for the normalized speed Ω_{PCn}*

2.4.2.3. *Controlling machines with a controllable excitation flux*

While there is an abundant quantity of scientific contributions concerning the high-speed control of fixed excited synchronous machines (Schiferl and Lipo 1990; Soong and Miller 1994), only a few contributions are dedicated to the study of control laws for controllable flux machines (wound excitation machines and hybrid excited machines).

The possibility of controlling the excitation flux would improve the high-speed operation of some synchronous machines. We will study the contribution that comes from this by adopting the trichotomy used in section 2.4.2.2.5.

2.4.2.3.1. $L_{dn} < 1$

By controlling the excitation flux, it is possible to maintain the power at its maximum value ($P_n = 1$) above the speed Ω_{UPFn1} for these machines. On the other hand, below this speed, the torque cannot be increased by controlling the excitation flux.

Figure 2.33(a) shows the maximum normalized power variation as a function of speed when the excitation flux is fixed $k_f = 1$ (a discontinuous curve), and when it is controllable, $k_f \in [0, 1]$ (a continuous curve), for a machine with ($L_{dn} = 0.8$, $\rho = 1.5$). Figure 2.33(b) illustrates the case where the optimal control strategy is applied to this machine in the (i_{dn}, i_{qn}) plane.

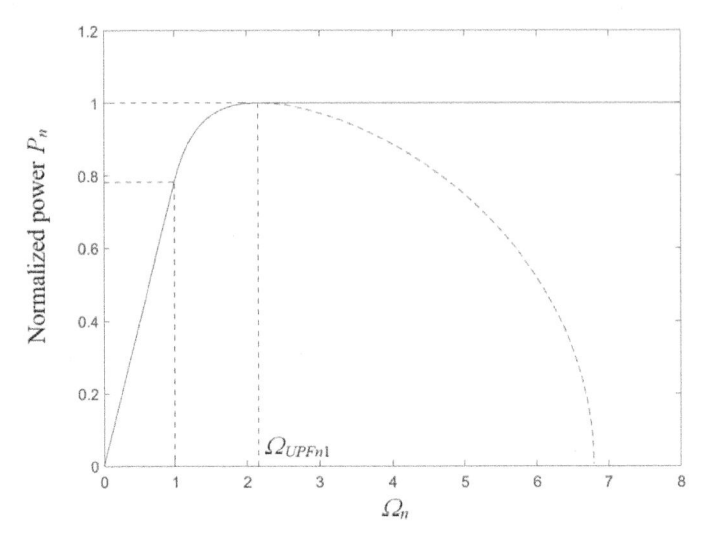

a) Normalized power variation as a function of speed ($L_{dn} = 0.8$, $\rho = 1.5$)

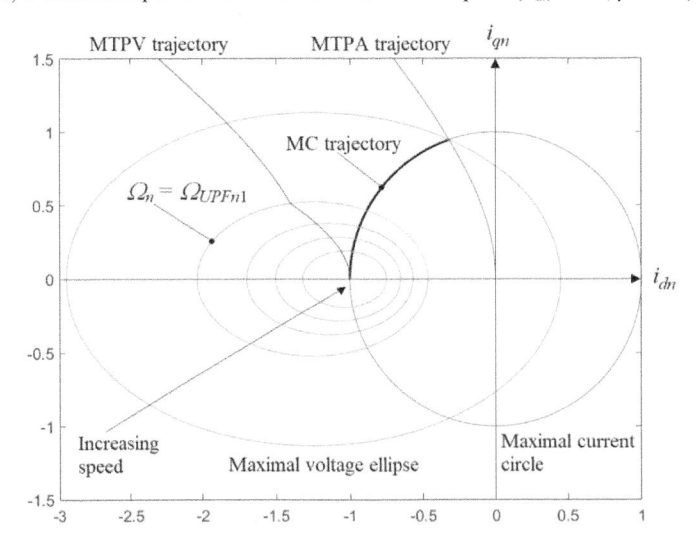

b) Representing the control strategy in the (i_{dn}, i_{qn}) plane

Figure 2.33. *Maximum normalized power variation of speed ($L_{dn} = 0.8$, $\rho = 1.5$) (a), and a representation of the control strategy in the (i_{dn}, i_{qn}) plane (b)*

The optimal control strategy is to adopt the MC control law. Controlling the excitation flux allows for the power $P_n = 1$ to be maintained when $\Omega_n \geq \Omega_{UPFn1}$. We can show that the excitation coefficient k_f is in this case given by (see the appendix: https://1drv.ms/b/s!AogSAGtYvycUkHIbc3FbsPoSQjJt?e=U9p85C):

$$k_f = \begin{cases} 1 & \Omega_n < \Omega_{UPFn1} \\ \dfrac{V_{n\,max}^2 + \rho \cdot (L_{dn} \cdot \Omega_n)^2}{\Omega_n \cdot \sqrt{V_{n\,max}^2 + (\rho \cdot L_{dn} \cdot \Omega_n)^2}} & \Omega_n \geq \Omega_{UPFn1} \end{cases} \quad [2.73]$$

Figure 2.34 presents how this coefficient varies as a function of the normalized speed, and we see that it tends towards the value of L_{dn} when the speed tends to infinity. This corresponds to the center of the boundary voltage ellipse being displaced towards the point (-1, 0) in the (i_{dn}, i_{qn}) plane. The operating range at a constant power is infinite in this case, and it suffices to have $k_{f\,min} = L_{dn}$.

Therefore, the optimal control strategy is to combine the control of the excitation and the armature currents for speeds $\Omega_n \geq \Omega_{UPFn1}$. The angle ψ corresponding to this control strategy was given by equation [2.74]. The mathematical developments allowing us to obtain the different quantities are available in the appendix (https://1drv.ms/b/s!AogSAGtYvycUkHIbc3FbsPoSQjJt?e=U9p85C).

$$\psi_{MC} = \begin{cases} \arcsin\left(\dfrac{L_{dn}^2 + k_f^2 - \left(V_{n\,max}^2 / \Omega_n^2\right)}{2 \cdot k_f \cdot L_{dn}} \right) & \rho = 1 \\ \arcsin \dfrac{k_f - \sqrt{\rho^2 \cdot \left(k_f^2 + (\rho^2 - 1) \cdot L_{dn}^2\right) + (1 - \rho^2) \cdot \left(\dfrac{V_{n\,max}}{\Omega_n}\right)^2}}{L_{dn} \cdot (1 - \rho^2)} & \rho \neq 1 \end{cases} \quad [2.74]$$

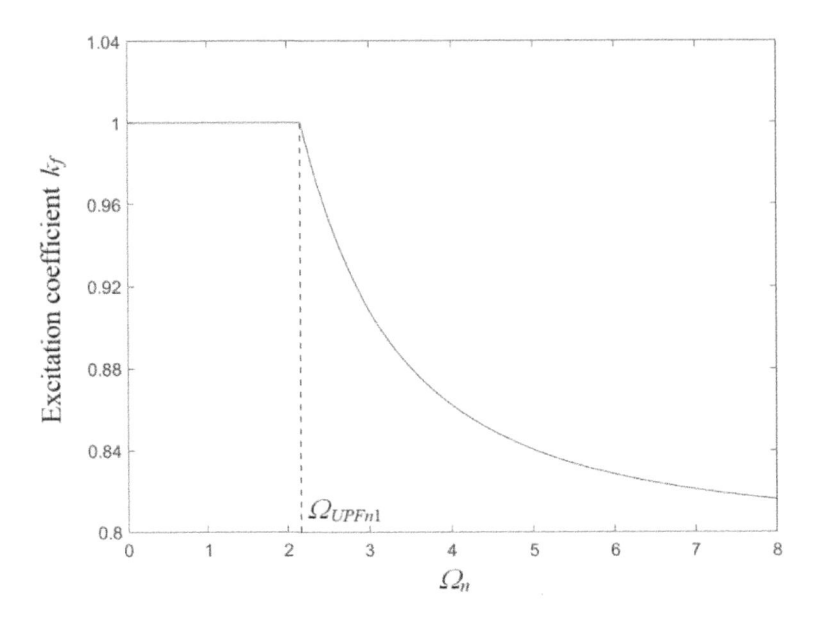

Figure 2.34. *Variation of the excitation coefficient k_f as a function of the normalized speed*

2.4.2.3.2. $L_{dn} = 1$

For machines with $\rho \geq 0.5$, controlling the excitation flux does not contribute towards any improvements. For machines with $\rho < 0.5$, controlling the flux allows for the normalized power to be maintained at its maximum value ($P_n = 1$) beyond the speed Ω_{UPFn1} [2.68]. The optimal control strategy to adopt is identical to the one defined previously for $L_{dn} < 1$.

2.4.2.3.3. $L_{dn} > 1$

For these machines, controlling the excitation flux only brings about improvements for structures whose parameters (L_{dn}, ρ) belong to "Zone 2" (Figure 2.28).

By controlling the excitation flux, the power maintains its maximum value ($P_n = 1$) when $\Omega_n \in [\Omega_{UPFn1}, \Omega_{UPFn2}]$. Figure 2.35(a) shows how the normalized power varies as a function of speed for a machine belonging to "Zone 2", with ($L_{dn} = 1.1$, $\rho = 0.15$), when the excitation flux is kept constant and equal to its maximum value ($k_f = 1$) (a discontinuous curve), and when the excitation flux can be controlled (a continuous curve), with:

$$k_f = \begin{cases} 1 & (\Omega_n < \Omega_{UPFn1}) \cup (\Omega_n > \Omega_{UPFn2}) \\ \dfrac{V_{n\,\max}^2 + \rho \cdot (L_{dn} \cdot \Omega_n)^2}{\Omega_n \cdot \sqrt{V_{n\,\max}^2 + (\rho \cdot L_{dn} \cdot \Omega_n)^2}} & \Omega_{UPFn1} \leq \Omega_n \leq \Omega_{UPFn2} \end{cases} \quad [2.75]$$

Figure 2.35(b) shows the variation of the excitation coefficient k_f as a function of speed.

Figure 2.36 shows the sectorization of "Zone 2", based on the contribution that controlling the excitation flux provides. This is subdivided into four sub-zones: "Zone 21", "Zone 22", "Zone 23" and finally "Zone 24".

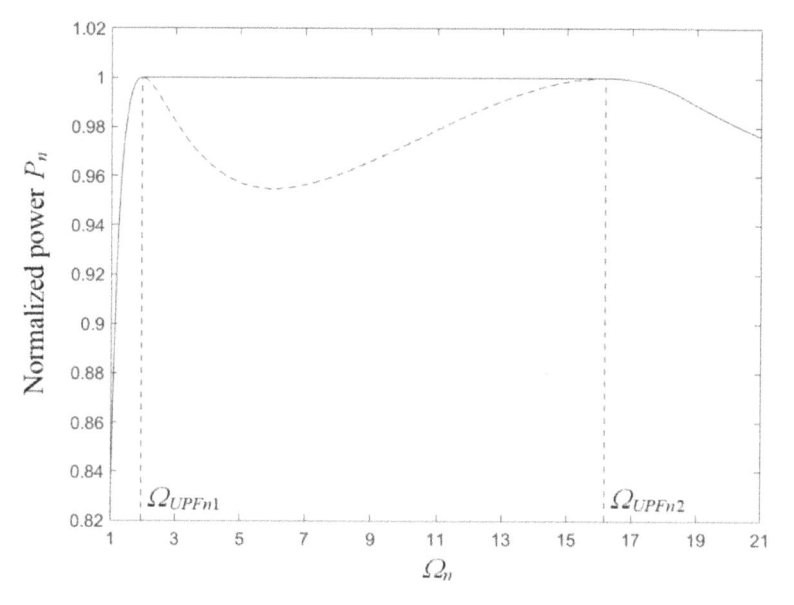

a) Variation of the normalized power as a function of speed ($L_{dn} = 1.1$, $\rho = 0.15$)

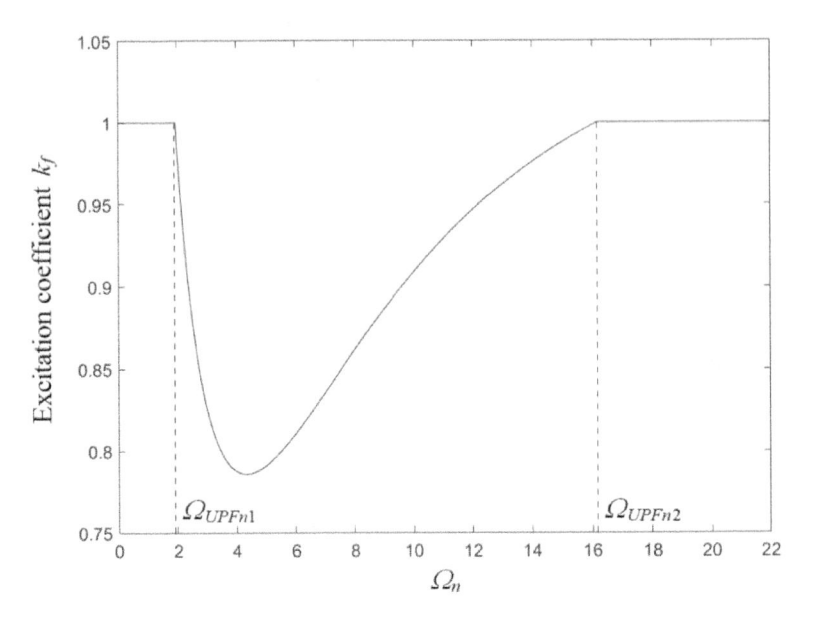

b) Variation of the excitation coefficient k_f as a function of normalized speed

Figure 2.35. *Characteristics of a machine with (L_{dn} = 1.1, ρ = 0.15), whose excitation flux is controlled [2.75] and uncontrolled (k_f = 1)*

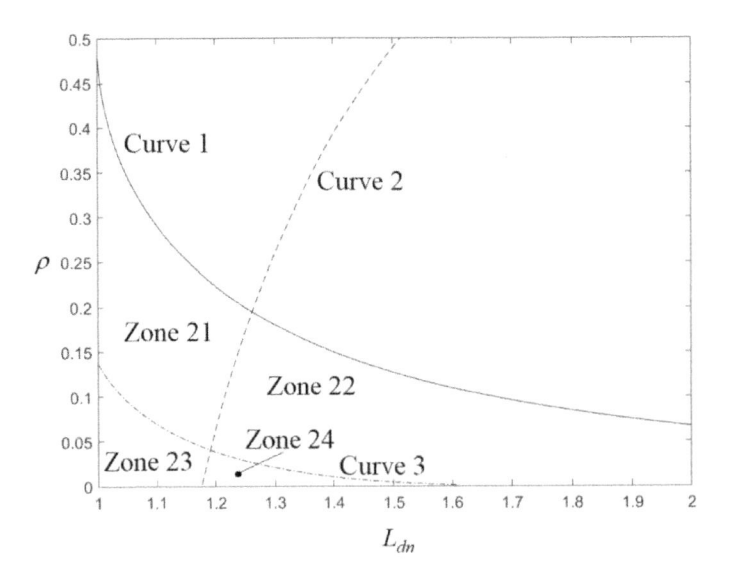

Figure 2.36. *Sectorization of "Zone 2"*

– "Zone 21" and "Zone 22": the speed Ω_{PCn} is not modified by controlling the excitation flux;

– "Zone 23": the speed $\Omega_{PCn} = +\infty$, due to the contribution coming from controlling the excitation flux;

– "Zone 24": the speed Ω_{PCn} increases, due to controlling the excitation flux, but remains finite.

The boundaries between these areas are defined by:

$$\begin{cases} \text{Curve 1:} & \text{Equations [2.58] and [2.69]} \\ \text{Curve 2:} & P_n(\Omega_n = 1) = \dfrac{1}{L_{dn}} = P_n(\Omega_n \to +\infty) \\ \text{Curve 3:} & P_n(\Omega_n = 1) = P_n(\Omega_{Opt1nMC}) \end{cases}$$

Therefore, by controlling the excitation flux of a synchronous machine, the operating range of their speeds can be extended. This contribution is particularly significant for machines with a relatively low d-axis armature magnetic reaction $L_{dn} < 1$.

2.4.2.4. Control of variable-reluctance synchronous machines

For variable-reluctance synchronous machines, it is not possible to operate it with a unitary power factor. Indeed, the relationship between the angles ψ and δ is given by:

$$\text{tg}\,\psi = \frac{-\rho}{\text{tg}(\delta)} \qquad [2.76]$$

Beyond the base speed, which can be deduced from equation [2.40]:

$$\Omega_b = \frac{\sqrt{2} \cdot V_{max}}{p \cdot I_{max} \cdot L_d \cdot \sqrt{1+\rho^2}} \qquad [2.77]$$

The three control laws that we studied for excited machines, namely the MTPA, the MTPV and the MC laws, can be adopted. We will study the characteristics of these laws for these machines at high-speed operations ($\Omega_n \geq 1$).

2.4.2.4.1. Maximum torque per ampere (MTPA) operation

This involves imposing the angle ψ given by equations [2.40], while ensuring that:

$$1 = \Omega_n \cdot \sqrt{\left(\rho \cdot L_{dn} \cdot I_n \cdot \cos(\psi_{opt})\right)^2 + \left(L_{dn} \cdot I_n \cdot \sin(\psi_{opt})\right)^2}$$

By solving this last equation, we obtain the variation in the armature current amplitude as a function of speed:

$$I_n = 1/\Omega_n \qquad [2.78]$$

The maximum speed, which can be obtained by adopting this control law, is infinite.

2.4.2.4.2. Maximum torque per volt (MTPV) operation

This is a question of imposing the angle δ maximizing the torque [2.16], which equals:

$$\delta_{opt} = 3 \cdot \pi/4 \qquad [2.79]$$

The expression for the normalized armature current amplitude is then given by:

$$I_n = \frac{1+\rho^2}{2 \cdot \rho \cdot \Omega_n} \qquad [2.80]$$

This amplitude is less than 1 only after passing a speed $\Omega_{\lim n}$, given by:

$$\Omega_{\lim n} = \frac{1+\rho^2}{2 \cdot \rho} \qquad [2.81]$$

Figure 2.37 shows how this speed varies as a function of the salience ratio ρ. The maximum operating speed that can be achieved by adopting this control law is infinite.

2.4.2.4.3. Maximum current operation (MC)

This control law aims to maintain the armature current amplitude at its maximum value $I_n = 1$ (Kron 1930; Terman et al. 1930; Sneyers 1985). To determine the angle ψ which should be fixed, the following equation must be solved:

$$1 = \Omega_n \cdot \sqrt{(\rho \cdot L_{dn} \cdot \cos(\psi))^2 + (L_{dn} \cdot \sin(\psi))^2}$$

Then, the angle ψ is the solution to the equation given by:

$$\psi_{MC} = \arccos\left(\sqrt{\frac{1}{(\rho^2-1)} \cdot \left(\frac{(1+\rho^2)}{2 \cdot \Omega_n^2} - 1\right)}\right) \qquad [2.82]$$

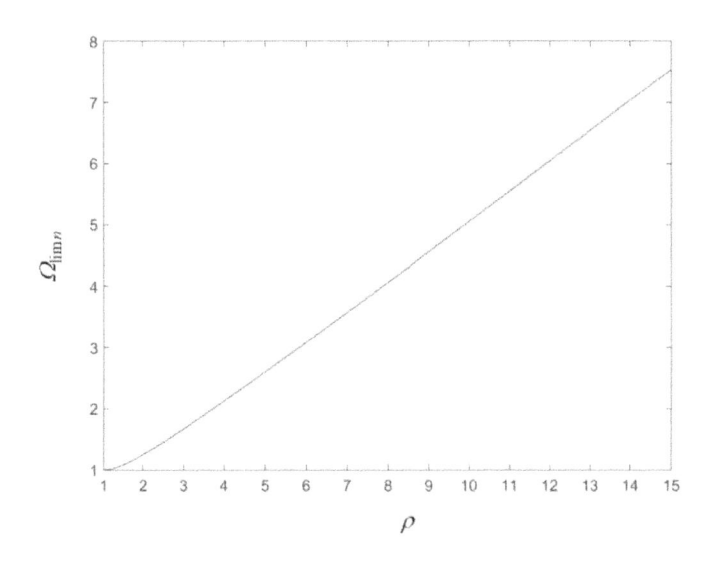

Figure 2.37. *Variation of speed $\Omega_{\lim n}$ as a function of the saliency ratio ρ*

The maximum speed that can be achieved by adopting this control law is given by:

$$\Omega_{nMC\,\max} = \sqrt{\frac{1+\rho^2}{2}} \qquad [2.83]$$

The torque, and therefore the power, cancel at this speed, and Figure 2.38 shows the variation of this speed as a function of the salience ratio ρ.

2.4.2.4.4. Optimal control strategy

For low speeds ($\Omega \leq \Omega_b \Leftrightarrow \Omega_n \leq 1$), the control law maximizing the torque while respecting current and voltage limits is the MTPA law.

At high speeds ($\Omega > \Omega_b \Leftrightarrow \Omega_n > 1$), the optimal control strategy that should be adopted in order to maximize the torque, while respecting the constraints ($V \leq V_{max}$ and $I \leq I_{max}$), first consists of imposing the MC control law until the speed $\Omega_{\lim n}$, and then imposing the MTPV control law afterwards. It can be shown that the MC control law produces a greater torque than the MTPA control law, when $\Omega_n \in (1, \Omega_{\lim n}]$ (see the appendix: https://1drv.ms/b/s!AogSAGtYvycUkHIbc3FbsPoSQjJt?e=U9p85C).

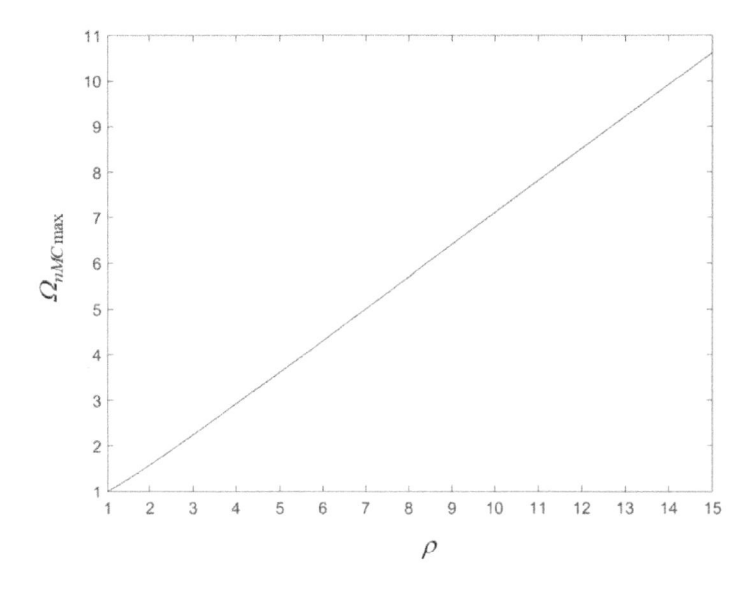

Figure 2.38. *Variation of speed Ω_{nMCmax} as a function of the saliency ratio ρ*

Figure 2.39(a) shows the maximum normalized power variation as a function of the normalized speed for a machine with $\rho = 8$. Figure 2.39(b) presents the optimal control strategy applied to this machine in the (i_{dn}, i_{qn}) plane. Note that, in the case of variable-reluctance synchronous machines which are not excited, the center of the boundary voltage ellipse is located at the origin ($i_{dn} = 0$, $i_{qn} = 0$).

The operating range for which it is possible to operate at a constant power corresponds to $\Omega_n \in [1, \Omega_{PCn}]$, and it can be shown that (see the appendix: https://1drv.ms/b/s!AogSAGtYvycUkHIbc3FbsPoSQjJt?e=U9p85C):

$$\Omega_{PCn} = \Omega_{\lim n} = \frac{1+\rho^2}{2 \cdot \rho} \qquad [2.84]$$

The larger the salience ratio, the larger the operating range at constant power (Figure 2.37).

Note that the power has an optimum (maximum) for the normalized speed Ω_{OptnMC} (Figure 2.39(a)), which is given by (see the appendix: https://1drv.ms/b/s!AogSAGtYvycUkHIbc3FbsPoSQjJt?e=U9p85C):

$$\Omega_{OptnMC} = \sqrt{\frac{1+\rho^2}{2 \cdot \rho}} \qquad [2.85]$$

This optimum occurs when the MC control law is applied. Figure 2.40 shows this variation in the speed as a function of the saliency ratio ρ.

2.5. Operations on the entire "torque/speed" plane

Section 2.4 was devoted to studying the control laws which allow for the "torque/speed" envelope to be optimized. It can be shown that, by controlling the excitation flux, we can improve the operation and enlarge this envelope. However, this fact is not unique to hybrid excited machines, which share this characteristic with purely wound excitation machines as well.

In this section, we will study the operation of synchronous machines on the entire "torque/speed" plane. This study is essential in order to formulate reliable conclusions that concern the dimensioning of hybrid excited machines. Indeed, for some applications (Figure 2.4) such as automotive electrical traction, partial-load operations are very important from an energy perspective, and it is therefore necessary to consider the losses in this study. We will see that hybrid excited machines have an additional degree of freedom (the hybridization ratio α) compared to other synchronous machines, therefore allowing the operation to have its energy optimized.

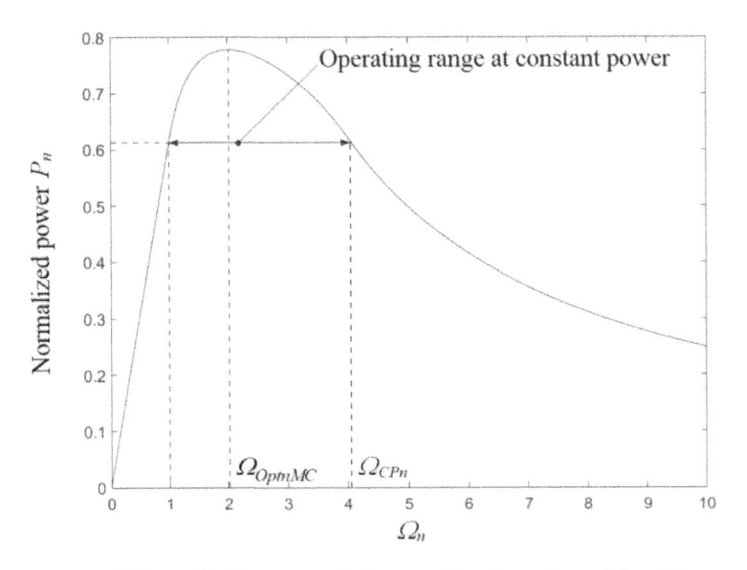

a) Normalized power variation as a function of speed ($\rho = 8$)

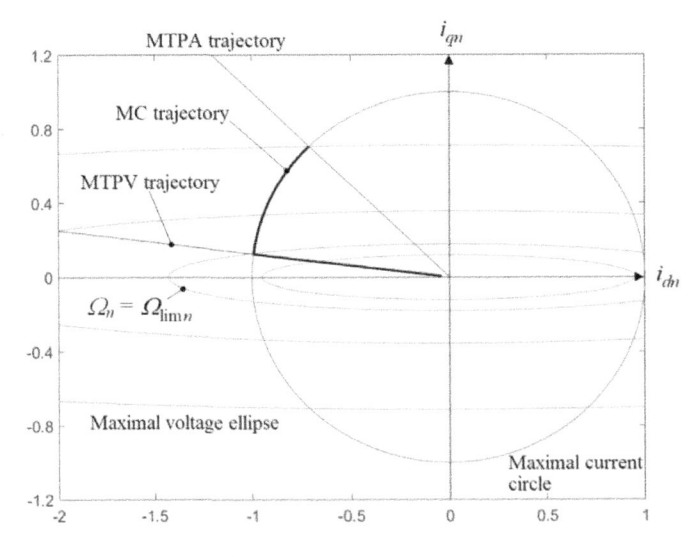

b) Representing the control strategy in the (i_{dn}, i_{qn}) plane

Figure 2.39. *Variation of maximum normalized power as a function of speed ($\rho = 8$) (a), and a representation of the control strategy in the (i_{dn}, i_{qn}) plane (b)*

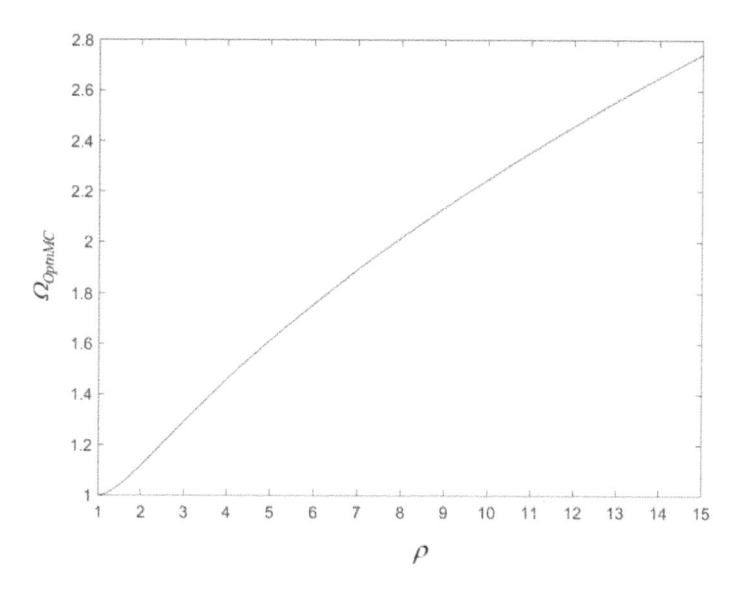

Figure 2.40. *Speed variation Ω_{OptnMC} as a function of the saliency ratio ρ*

Here, we will describe how algorithms can be implemented, allowing us to determine which control strategies optimize the energy consumption (efficiency optimization) on the entire "torque/speed" operating plane. These algorithms will show the contribution that hybrid excitation provides compared to other synchronous machines. The aim is to study the contributions coming from hybrid excitation, as well as to provide the readers with the tools that will allow them to carry out their own parametric studies.

2.5.1. *Efficiency optimization algorithms on the entire "torque/ speed" plane*

In this section, we will introduce the basic principles that will allow us to analyze how synchronous machines operate on the entire "torque/speed" plane, and also how to dimension them appropriately by taking their complete operating cycle into account.

The core of these algorithms is based on maximizing the efficiency, and consequently on minimizing the losses for each considered "torque/speed" operating point. The objective function is defined by:

$$f(\underbrace{I, \psi, k_f}_{\text{Control}}, \underbrace{\alpha, \beta, L_d, \rho}_{\text{Design}}) = P_{Cu} + P_{Fe} = R_s \cdot I^2 + R_e \cdot I_e^2 + R_f \cdot I_f^2 \qquad [2.86]$$

For a given "torque/speed" operating point, the total losses are a function of two types of quantities [2.86]: one type corresponds to control quantities and the other type is related to the design.

2.5.1.1. *Algorithms to study operations on the entire "torque/speed" plane*

When analyzing the operations on the entire "torque/speed" plane, we will assume that the quantities related to the design are fixed. The algorithm can then be used to determine the control variables, which lead to optimized efficiency for the machine in question. Figure 2.41 presents the algorithm developed for this purpose and can be used to generate "look-up tables" for control purposes. Two quantities are important when establishing this algorithm:

– the calculation of the efficiency η;

– the calculation of the normalized voltage limit value $V_{n\,\max}$.

Depending on whether the machine has non-salient or salient poles, it is possible to determine certain quantities that are useful when establishing this algorithm in a purely analytical or numerical way, respectively. We will cover these aspects in detail in sections 2.5.2 and 2.5.3.

The first step that needs to be taken when constructing this algorithm, as well as those developed for the sizing, concerns aspects common to both types of machines ($\rho = 1$ and $\rho \neq 1$), and also to all synchronous machines more generally. Given the normalized quantities defined and adopted in section 2.4.1.2, in particular the excitation flux equation [2.36], and by considering the excitation coefficient $k_f \in [0, 1]$, it can be shown that the quantities k_{en} and β depend on the hybridization ratio α, and that we have (see the appendix: https://1drv.ms/b/s!AogSAGtYvycUkHIbc3FbsPoSQjJt?e=U9p85C):

$$\begin{cases} k_{en} = \begin{cases} \alpha & \alpha \in [0.5, 1] \\ (1-\alpha) & \alpha \in [0, 0.5) \end{cases} \\ \beta = \begin{cases} \beta_1/\alpha^2 & \alpha \in [0.5, 1] \\ \beta_1/(1-\alpha)^2 & \alpha \in [0, 0.5) \end{cases} \end{cases} \qquad [2.87]$$

where β_1 is a reference value that corresponds to the value of β, when the wound excitation is used only to weaken the excitation flux in hybrid excited machines, i.e. when $\alpha = 1$.

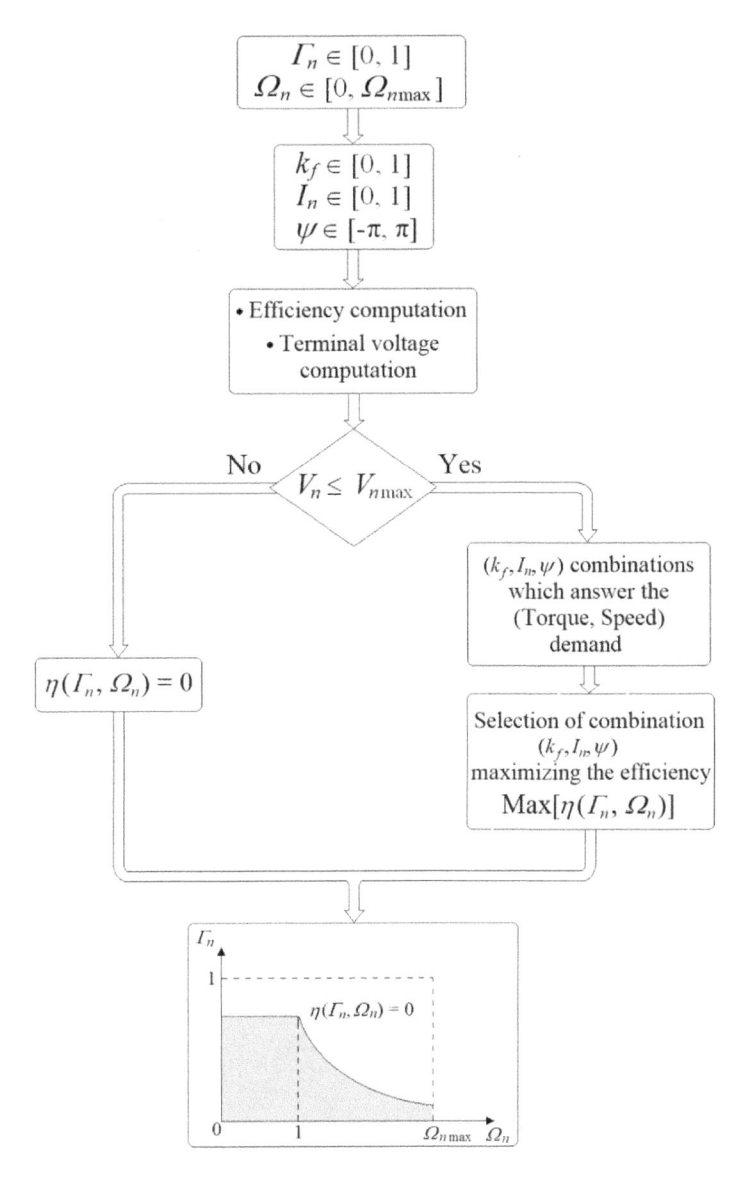

Figure 2.41. *The algorithm used to calculate the efficiency mapping in the "torque/speed" plane*

The efficiency is given by:

$$\eta = \frac{\Gamma_{em} \cdot \Omega}{\Gamma_{em} \cdot \Omega + R_s \cdot I^2 + R_e \cdot I_e^2 + R_f \cdot I_f^2} \qquad [2.88]$$

The normalized voltage limit $V_{n\,max}$ is calculated using the algorithm presented in Figure 2.42.

Establishing the interdependence of the parameters α, β and k_{en}, as well as the efficiency calculation and the value of $V_{n\,max}$, makes up the first step to be taken, and also forms the core of the developed algorithms.

2.5.1.2. *Sizing algorithms*

For sizing purposes, quantities that come up when designing synchronous machines are considered to be optimization variables. The aim here is still the same as in [2.86].

Figure 2.43 presents an algorithm developed with the aim of optimizing the hybridization ratio of a hybrid excited synchronous machine for a particular operating point in the "torque/speed" plane (Amara et al. 2021a). By adopting the same approach, it is possible to develop a more general algorithm which includes each dimensional parameter, i.e. α, β_1, L_d and ρ.

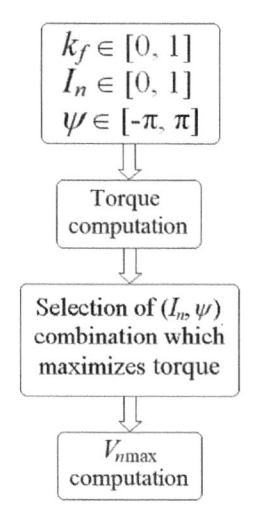

$k_f \in [0, 1]$
$I_n \in [0, 1]$
$\psi \in [-\pi, \pi]$

↓

Torque computation

↓

Selection of (I_n, ψ) combination which maximizes torque

↓

V_{nmax} computation

Figure 2.42. *The algorithm used to calculate $V_{n\,max}$*

Control of Hybrid Excited Synchronous Machines 117

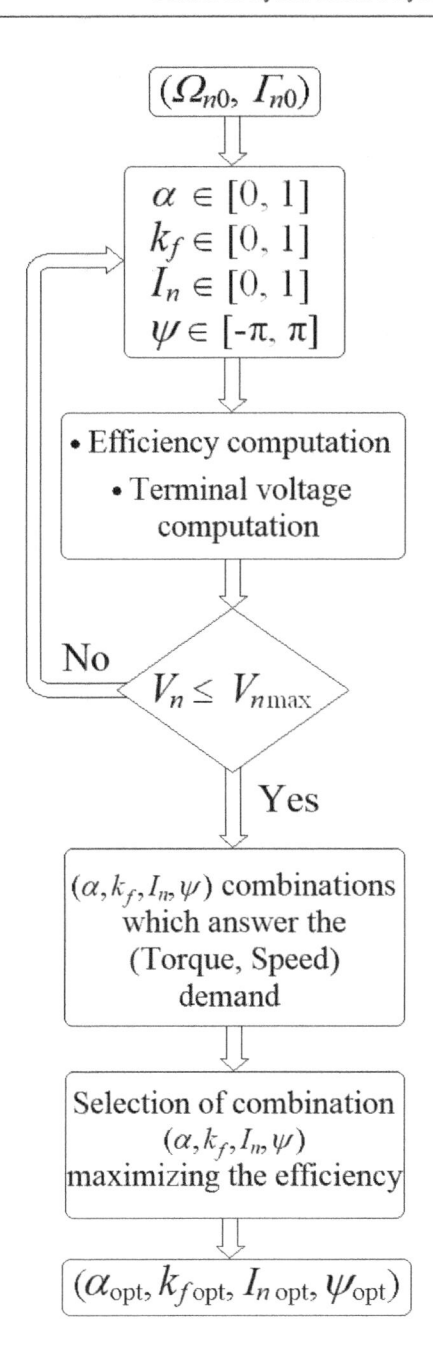

$$(\Omega_{n0},\ \Gamma_{n0})$$

$\alpha \in [0, 1]$
$k_f \in [0, 1]$
$I_n \in [0, 1]$
$\psi \in [-\pi, \pi]$

- Efficiency computation
- Terminal voltage computation

$V_n \leq V_{n\max}$? — No (loop back)

Yes

(α, k_f, I_n, ψ) combinations which answer the (Torque, Speed) demand

Selection of combination (α, k_f, I_n, ψ) maximizing the efficiency

$$(\alpha_{\mathrm{opt}},\ k_{f\mathrm{opt}},\ I_{n\,\mathrm{opt}},\ \psi_{\mathrm{opt}})$$

Figure 2.43. *The algorithm used to optimize the hybridization ratio α*

In the following section, the algorithms presented in Figures 2.41, 2.42 and 2.43 are discussed first in the case of synchronous machines with non-salient poles. This will allow us to clearly formulate the steps leading to the efficiency map calculations in the "torque/speed" plane.

For non-salient pole machines ($\rho = 1$), the number of steps that can be analytically conducted is greater than that for salient pole machines ($\rho \neq 1$). Therefore, we will use two different approaches for these two different types of machines.

Since non-salient pole machines are a special case of salient pole ones, this will provide us with two tools based upon two different approaches for non-salient pole machines.

The whole point of having these two different tools is so that we can compare the two approaches and therefore verify their accuracy. Indeed, for the same machine, the two different tools should give the same results. Moreover, previous studies of the control laws based on the lossless model can also be used as references. The model with losses is equivalent to the lossless model by imposing the following: $R_e = 0\ \Omega$, $R_s = 0\ \Omega$ and $R_f = +\infty$.

2.5.2. *Normalized model with losses and the calculation of $V_{n\,max}$*

The iron loss current I_f and the current I_0 (Figure 2.1) are both normalized with respect to the maximum armature current magnitude I_{max}. The voltage V_0 (Figure 2.1) is normalized with respect to the EMF (electromotive force) at the base speed Ω_b, and so is the armature voltage [2.30].

In doing this, equations [2.1] and [2.2] can be rewritten using the normalized quantities as follows:

$$\begin{bmatrix} v_{dn} \\ v_{qn} \end{bmatrix} = R_{ns} \cdot \begin{bmatrix} i_{dn} \\ i_{qn} \end{bmatrix} + \begin{bmatrix} v_{0dn} \\ v_{0qn} \end{bmatrix}$$

$$\begin{bmatrix} v_{0dn} \\ v_{0qn} \end{bmatrix} = \begin{bmatrix} 0 & -\Omega_n \cdot \rho \cdot L_{dn} \\ \Omega_n \cdot L_{dn} & 0 \end{bmatrix} \cdot \begin{bmatrix} i_{0dn} \\ i_{0qn} \end{bmatrix} + \Omega_n \cdot k_f \cdot \begin{bmatrix} 0 \\ 1 \end{bmatrix}$$

$$\begin{bmatrix} i_{dn} \\ i_{qn} \end{bmatrix} = \begin{bmatrix} i_{fdn} \\ i_{fqn} \end{bmatrix} + \begin{bmatrix} i_{0dn} \\ i_{0qn} \end{bmatrix}$$

$$\begin{bmatrix} i_{fdn} \\ i_{fqn} \end{bmatrix} = \frac{1}{R_{fn}} \cdot \left(\begin{bmatrix} 0 & -\Omega_n \cdot \rho \cdot L_{dn} \\ \Omega_n \cdot L_{dn} & 0 \end{bmatrix} \cdot \begin{bmatrix} i_{0dn} \\ i_{0qn} \end{bmatrix} + \Omega_n \cdot k_f \cdot \begin{bmatrix} 0 \\ 1 \end{bmatrix} \right)$$

[2.89]

The normalized losses are then given by:

$$\begin{cases} P_{Cu\,n} = \dfrac{P_{Cu}}{V_{max} \cdot I_{max}} = \dfrac{R_{sn} \cdot I_n^2}{V_{n\,max}} + \dfrac{R_{en} \cdot (k_f - \alpha)^2}{\beta \cdot k_{en}^2} & \text{(Joule losses)} \\[2mm] P_{Fe\,n} = \dfrac{P_{Fe}}{V_{max} \cdot I_{max}} = \dfrac{R_{fn} \cdot I_{fn}^2}{V_{n\,max}} & \text{(iron losses)} \end{cases}$$

[2.90]

and the normalized torque by:

$$\Gamma_n = \frac{i_{0qn}}{V_{n\,max}} \cdot \left(k_f + (1-\rho) \cdot L_{dn} \cdot i_{0dn} \right)$$

[2.91]

From equation [2.89], it is possible to express the components i_{0dn} and i_{0qn} of the normalized current I_0 in terms of the components i_{dn} and i_{qn} of the normalized armature current I_n:

$$\begin{bmatrix} i_{0dn} \\ i_{0qn} \end{bmatrix} = \frac{1}{1 + \dfrac{\rho \cdot (L_{dn} \cdot \Omega_n)^2}{R_{fn}^2}} \cdot \begin{bmatrix} 1 & \dfrac{\Omega_n \cdot \rho \cdot L_{dn}}{R_{fn}} \\ -\dfrac{\Omega_n \cdot L_{dn}}{R_{fn}} & 1 \end{bmatrix} \cdot \left(\begin{bmatrix} i_{dn} \\ i_{qn} \end{bmatrix} - \begin{bmatrix} 0 \\ \dfrac{\Omega_n \cdot k_f}{R_{fn}} \end{bmatrix} \right)$$

[2.92]

We need equations [2.91] and [2.92] to calculate the normalized voltage $V_{n\,max}$. Figure 2.42 details the computer code used in this calculation.

The functions f and g (Figure 2.42) are given in equation [2.92]. The function h is deduced from the normalized torque equation [2.91] and is given by:

$$h = \Gamma_n \cdot V_{n\ max} = i_{0qn} \cdot (1 + (1-\rho) \cdot L_{dn} \cdot i_{0dn})$$ [2.93]

For a given machine, maximizing the function h corresponds to maximizing the torque. Note that k_f is set equal to 1 in this function since the maximum torque is obtained by imposing the maximum excitation flux (see sections 2.3.1 and 2.3.2).

Start
Definition of the parameters ρ, L_{dn}, R_{sn}, R_{fn}.
$I_n = [0 : not_I : 1]$;
$\psi = [-\pi : not_\psi : \pi]$;
For $i = 1$: length(I_n)
    For $j = 1$: length(ψ)
        $i_{dn}(j) = -I_n(i) \cdot \sin(\psi(j))$;
        $i_{qn}(j) = I_n(i) \cdot \cos(\psi(j))$;
        $i_{0dn}(j) = f(i_{dn}(j), i_{qn}(j))$;
        $i_{0qn}(j) = g(i_{dn}(j), i_{qn}(j))$;
        $\Gamma(j) = h(i_{0dn}(j), i_{0qn}(j))$;
    End
    $[Y1(i), X1(i)] = \max(\Gamma)$;
    $\psi_1(i) = \psi(X1(i))$;
End
$[Y2, X2] = \max(Y1)$;
$I_{Opt} = I_n(X2)$;
$\psi_{Opt} = \psi_1(X2)$;
$i_{dnOpt} = -I_{Opt} \cdot \sin(\psi_{Opt})$;
$i_{qnOpt} = I_{Opt} \cdot \cos(\psi_{Opt})$;
$i_{0dnOpt} = f(i_{dnOpt}, i_{qnOpt})$;
$i_{0qnOpt} = g(i_{dnOpt}, i_{qnOpt})$;
$V_{n\ max} = v(i_{dnOpt}, i_{qnOpt}, i_{0dnOpt}, i_{0qnOpt})$;
End

Figure 2.44. *Script that calculates $V_{n\ max}$*

The definition of the voltage $V_{n\,max}$ corresponds to that of the base speed, which is the maximum achievable speed from maintaining the MTPA control law. Once the components of the current I_n that maximize the torque have been determined, it is possible to calculate the maximum normalized voltage $V_{n\,max}$. The function v (Figure 2.42) is deduced from equation [2.89]:

$$v = \sqrt{\left(R_{sn} \cdot i_{dn} - \Omega_n \cdot \rho \cdot L_{dn} \cdot i_{0qn}\right)^2 + \left(R_{sn} \cdot i_{qn} + \Omega_n \cdot L_{dn} \cdot i_{0dn} + 1\right)^2} \qquad [2.94]$$

After calculating the value of $V_{n\,max}$, let us describe how to implement the previous algorithms for synchronous machines with non-salient ($\rho = 1$) and salient poles ($\rho \neq 1$).

2.5.3. *Machines with non-salient poles ($\rho = 1$)*

For machines with non-salient poles, the formula for the torque (equation [2.93] with ($\rho = 1$)) is simpler and the same is true for the expression for the losses.

Here, maximizing the torque corresponds to maximizing the i_{0qn} component. This simplifies the calculation of the value of $V_{n\,max}$.

Without assuming that the current ($I_n \leq 1$) and voltage ($V_n \leq V_{n\,max}$) constraints are satisfied and, for a given operating point on the "torque/speed" plane and a given value of the excitation flux, it is possible to determine the relationship between the i_{0qn} component and the normalized torque Γ_n:

$$i_{0qn} = \frac{\Gamma_n \cdot V_{n\,max}}{k_f} \qquad [2.95]$$

For this operating point, it follows that the total losses can be expressed as just a function of the i_{0dn} component. The losses are in fact a quadratic function of the i-$_{0dn}$ component (see the appendix: https://1drv.ms/b/s!AogSAGtYvycUkHIbc3FbsPoSQjJt?e=U9p85C). It can then be shown that the value of this component which maximizes the efficiency [2.88] or, in other words, the one that minimizes the losses, is the solution of the following equation:

$$\frac{\partial(P_{Cu\,n} + P_{Fe\,n})}{\partial i_{0dn}} = 0 \qquad [2.96]$$

The value of i_{0dn}, which is the solution of equation [2.96], is given by (see the appendix: https://1drv.ms/b/s!AogSAGtYvycUkHIbc3FbsPoSQjJt?e=U9p85C) (Morimoto et al. 1994):

$$i_{0dOptn} = \frac{-\Omega_n^2 \cdot L_{dn} \cdot (R_{sn} + R_{fn}) \cdot k_f}{\left(R_{sn} \cdot R_{fn}^2 + \Omega_n^2 \cdot L_{dn}^2 \cdot (R_{sn} + R_{fn})\right)} \qquad [2.97]$$

Furthermore, it can be shown that the current ($I_n \leq 1$) and voltage ($V_n \leq V_{n\,max}$) constraints result in the following equations:

$$I_n \leq 1 \Leftrightarrow A_I \cdot i_{0dn}^2 + B_I \cdot i_{0dn} + C_I \leq 0 \qquad [2.98]$$

$$V_n \leq V_{n\,max} \Leftrightarrow A_V \cdot i_{0dn}^2 + B_V \cdot i_{0dn} + C_V \leq 0 \qquad [2.99]$$

where

$$\begin{cases} A_I = 1 + \left(\Omega_n \cdot L_{dn}/R_{fn}\right)^2 \\ B_I = 2 \cdot \left(\Omega_n^2 \cdot k_f \cdot L_{dn}/R_{fn}^2\right) \\ C_I = \left(\left(\Omega_n \cdot L_{dn} \cdot \Gamma_n \cdot V_{n\,max}\right)^2 + \left(\Gamma_n \cdot V_{n\,max} \cdot R_{fn} + \Omega_n \cdot k_f^2\right)^2\right) \big/ \left(k_f \cdot R_{fn}\right)^2 - 1 \end{cases}$$

and

$$\begin{cases} A_V = L_{dn}^2 \cdot \Omega_n^2 \cdot \left(1 + (R_{sn}/R_{fn})\right)^2 + R_{sn}^2 \\ B_V = 2 \cdot \Omega_n^2 \cdot k_f \cdot L_{dn} \cdot \left(1 + (R_{sn}/R_{fn})\right)^2 \\ C_V = \begin{bmatrix} \left(k_f \cdot \Omega_n \cdot (1 + R_{sn}/R_{fn}) + (R_{sn} \cdot \Gamma_n \cdot V_{n\,max}/k_f)\right)^2 \\ + \left(L_{dn} \cdot \Omega_n \cdot \Gamma_n \cdot V_{n\,max} \cdot (1 + R_{sn}/R_{fn})/k_f\right)^2 - V_{n\,max}^2 \end{bmatrix} \end{cases}$$

The algorithm described previously, which calculates the efficiency mappings (Figure 2.41), can be implemented for machines with non-salient poles from the last three equations [2.79], [2.98] and [2.99].

For a given machine, whose parameters α, β, k_{en}, R_{sn}, R_{fn} and L_{dn} are known and for which the value $V_{n\,max}$ is determined, for a given value of the excitation flux k_f and from a given starting operating point in the "torque/speed" plane, the combination (k_f, I_n, ψ) is selected by the following four steps (Figure 2.45):

1) compute the value of i_{0dOptn} [2.97];

2) determine the interval [i_{0dnI1}, i_{0dnI2}] for which the current constraint is satisfied ($I_n \leq 1$) from equation [2.98];

3) determine the interval [i_{0dnV1}, i_{0dnV2}] for which the current constraint is satisfied ($V_n \leq V_{n\,max}$) from equation [2.99];

4) select the combination (k_f, I_n, ψ) maximizing the efficiency (minimizing losses), while still meeting the current and voltage constraints (Figure 2.45).

Start
If the discriminant of equation [2.98] is negative $\Rightarrow \eta(\Gamma_n, \Omega_n, k_f) = 0$
Or if the discriminant of equation [2.99] is negative $\Rightarrow \eta(\Gamma_n, \Omega_n, k_f) = 0$
Otherwise
If [i_{0dnI1}, i_{0dnI2}] $\cap$ [i_{0dnV1}, i_{0dnV2}] = $\emptyset \Rightarrow \eta(\Gamma_n, \Omega_n, k_f) = 0$
Otherwise
If $i_{0dOptn} \in$ [i_{0dnI1}, i_{0dnI2}] $\cap$ [i_{0dnV1}, i_{0dnV2}] = $\emptyset \Rightarrow \eta = \eta(\Gamma_n, \Omega_n, k_f, i_{0dOptn})$
Otherwise $\eta = \eta(\Gamma_n, \Omega_n, k_f, i_{0dOptn1})$, where $i_{0dOptn1}$ is the bound of the interval [i_{0dnI1}, i_{0dnI2}] $\cap$ [i_{0dnV1}, i_{0dnV2}] closest to i_{0dOptn}.
End

Figure 2.45. *Script used to choose the combination (k_f, I_n, ψ) (inner loop)*

The first three steps are independent of one another, so they can be performed in either order. For the fourth step, the results of the previous three steps are required to make the selection. For a given operating point, if it is not possible to find a combination (k_f, I_n, ψ) satisfying the different criteria, i.e. maximizing the efficiency within the constraints, a zero value is chosen to be the efficiency $\eta(\Gamma_n, \Omega_n) = 0$. Figure 2.45 proposes the script implemented to perform the selection. The computer code developed to plot the efficiency mappings in the "torque/speed" plane contains three loops, and this script describes what is done in the inner loop (Amara et al. 2021a).

The three loops are nested (Amara 2001; Amara et al. 2021a): the first is for the speed variation interval Ω_n (first outer loop), the second is for the torque variation interval Γ_n (second outer loop) and the third is for the excitation coefficient variation interval k_f (inner loop). The script in Figure 2.45 is implemented in the inner loop to select the optimal combination (k_f, I_n, ψ).

The incremental loops of the speed Ω_n and of the torque Γ_n are both considered to be external loops because they are interchangeable. The "torque/speed" plane is in fact gridded by a regular mesh, the fineness of which depends on the variational steps in the normalized speed and torque intervals. The grid limits depend on the limits of the intervals.

On the inner loop level, the previous script selects the optimal pair (I_n, ψ), if any, maximizing the efficiency for each value of k_f of a given torque (Ω_n, Γ_n). If there is no pair (I_n, ψ) that maximizes the efficiency within the constraints, then the efficiency is forced to be zero ($\eta = 0$). The selection of the optimal triple (k_f, I_n, ψ) is done simply once the interval of k_f has been fully determined.

Solutions to the inequalities [2.98] and [2.99] are necessarily bounded intervals since A_I and A_V are both strictly positive. The values of i_{0dnI1} and i_{0dnI2} are given by:

$$i_{0dnI1} = \frac{-B_I - \sqrt{B_I^2 - 4 \cdot A_I \cdot C_I}}{2 \cdot A_I} \quad i_{0dnI2} = \frac{-B_I + \sqrt{B_I^2 - 4 \cdot A_I \cdot C_I}}{2 \cdot A_I}$$

and those of i_{0dnV1} and i_{0dnV2} by:

$$i_{0dnV1} = \frac{-B_V - \sqrt{B_V^2 - 4 \cdot A_V \cdot C_V}}{2 \cdot A_V} \quad i_{0dnV2} = \frac{-B_V + \sqrt{B_V^2 - 4 \cdot A_V \cdot C_V}}{2 \cdot A_V}$$

2.5.4. *Machines with salient poles ($\rho \neq 1$)*

For salient pole machines ($\rho \neq 1$), the computer code developed to plot the efficiency mappings in the "torque/speed" plane contains four loops: the first is for the speed variation interval Ω_n (first outer loop), the second is for the torque variation interval Γ_n (second outer loop), the third is for the excitation coefficient variation interval k_f (first inner loop) and the last is for the angle variation interval ψ (second inner loop).

There is an additional loop compared to the previous code developed for non-salient pole machines, and the number of steps that can be conducted analytically is equal to 1. This is a rewriting of the normalized torque equation given in [2.91], where the quantities Ω_n, Γ_n, k_f and ψ are considered to be parameters, in addition to the parameters ρ, L_{dn}, R_{fn} and $V_{n\,max}$ [2.100].

From the torque equation [2.91] and equation [2.92], it is possible to define a quadratic equation in the armature current amplitude I_n:

$$A_\Gamma \cdot I_n^2 + B_\Gamma \cdot I_n + C_\Gamma = 0 \qquad [2.100]$$

where

$$\begin{cases} A_\Gamma = L_{dn} \cdot (1-\rho) \cdot \left(\left(\Omega_n \cdot \rho \cdot L_{dn}/R_{fn} \right) \cdot \cos\psi - \sin\psi \right) \cdot \left(\left(\Omega_n \cdot L_{dn}/R_{fn} \right) \cdot \sin\psi + \cos\psi \right) \\[2mm] B_\Gamma = \begin{pmatrix} \left(\left(\Omega_n \cdot L_{dn}/R_{fn} \right) \cdot \sin\psi + \cos\psi \right) \cdot \begin{pmatrix} \left(1+\rho \cdot \left(\Omega_n \cdot L_{dn}/R_{fn} \right)^2 \right) \cdot k_f \\ -\left(\dfrac{\Omega_n^2 \cdot (1-\rho) \cdot \rho \cdot L_{dn}^2 \cdot k_f}{R_{fn}^2} \right) \end{pmatrix} \\ -\left(\Omega_n \cdot k_f/R_{fn} \right) \cdot L_{dn} \cdot (1-\rho) \cdot \left(\left(\Omega_n \cdot \rho \cdot L_{dn}/R_{fn} \right) \cdot \cos\psi - \sin\psi \right) \end{pmatrix} \\[2mm] C_\Gamma = \begin{pmatrix} -\left(\dfrac{\Omega_n \cdot k_f}{R_{fn}} \right) \cdot \left(\left(1+\rho \cdot \left(\Omega_n \cdot L_{dn}/R_{fn} \right)^2 \right) \cdot k_f - \left(\dfrac{\Omega_n^2 \cdot (1-\rho) \cdot \rho \cdot L_{dn}^2 \cdot k_f}{R_{fn}^2} \right) \right) \\ -\Gamma_n \cdot \left(1+\rho \cdot \left(\Omega_n \cdot L_{dn}/R_{fn} \right)^2 \right)^2 \cdot V_{n\,max} \end{pmatrix} \end{cases}$$

The two external or the two internal loops are both interchangeable. It should be noted that the internal loop concerning the excitation coefficient k_f

can be replaced exactly by a loop for the normalized amplitude of the armature current I_n. Indeed, equation [2.100] can be easily rewritten, leading to an equation quadratic in the excitation coefficient k_f.

It is important to maintain an inner loop that includes the angle ψ because rewriting equation [2.100] with ψ as an unknown leads to a formulation whose analytical solution seems very complicated. It is therefore more appropriate to maintain a purely analytical treatment of this quantity.

Figure 2.46 presents the script implemented, which selects a combination (k_f, I_n, ψ) that maximizes the efficiency, while still satisfying the current and voltage constraints. As before, if, for a given operating point, it is not possible to find such a combination (k_f, I_n, ψ) that satisfies the various criteria, then the zero value is chosen for the efficiency $\eta\,(\Gamma_n, \Omega_n) = 0$.

Solutions of equation [2.100] are the values of I_{n1} and I_{n2} given by:

$$\begin{cases} I_{n1} = \dfrac{-B_\Gamma - \sqrt{B_\Gamma^2 - 4 \cdot A_\Gamma \cdot C_\Gamma}}{2 \cdot A_\Gamma} \\[2ex] I_{n2} = \dfrac{-B_\Gamma + \sqrt{B_\Gamma^2 - 4 \cdot A_\Gamma \cdot C_\Gamma}}{2 \cdot A_\Gamma} \end{cases} \quad [2.101]$$

Start
If the discriminant of equation [2.100] is negative $\Rightarrow \eta(\Gamma_n, \Omega_n, k_f, \psi, I_n) = 0$
Otherwise
Equation [2.100] has two solutions: I_{n1} and I_{n2}. We then have three cases:
    1) both solutions $I_n > 1 \Rightarrow \eta(\Gamma_n, \Omega_n, k_f, \psi, I_n) = 0$;
    2) one of these two solutions respects the current constraint
        $\Rightarrow \eta = \eta(\Gamma_n, \Omega_n, k_f, \psi, I_n)$, where I_n is the solution respecting the current constraint;
    3) both solutions respect the current constraint
        $\Rightarrow \eta = \max[\eta\,(\Gamma_n, \Omega_n, k_f, \psi, I_{n1}), \eta\,(\Gamma_n, \Omega_n, k_f, \psi, I_{n2})]$.
End

Figure 2.46. *Script for selecting a combination (k_f, I_n, ψ) (inner loop)*

Once the two intervals of k_f and ψ have been completely determined, the two inner loops allow for an optimizing triple (k_f, I_n, ψ) to be chosen which, for a given torque (Ω_n, Γ_n), allows for the efficiency to be maximized. As before, if there is no triple (k_f, I_n, ψ) that maximizes the efficiency within the constraints, then the efficiency is set to zero ($\eta = 0$).

In the special case of non-salient pole machines ($\rho = 1$), equation [2.100] becomes:

$$B_\Gamma \cdot I_n + C_\Gamma = 0 \qquad [2.102]$$

and the value of the normalized armature current amplitude I_n is given by:

$$I_n = \frac{-C_\Gamma}{B_\Gamma} \qquad [2.103]$$

This does not fundamentally change the script presented in Figure 2.46, and even allows for it to be simplified since then there would no longer be two solutions to evaluate, but only one.

2.5.4.1. *The case of variable-reluctance synchronous machines*

Even if strong similarities exist within the efficiency mapping calculations on the entire "torque/speed" plane, for any synchronous machine, whether it is excited (permanent magnet, wound excitation and hybrid excited machines) or not (variable-reluctance synchronous machines), there are differences that can be identified from a global mathematical point of view.

These differences are related to the lack of an excitation flux for variable-reluctance synchronous machines, resulting in different normalization systems being used for excited and variable-reluctance synchronous machines, as discussed in sections 2.4.1.2 and 2.4.1.3. We will discuss these differences in the following section.

2.5.4.1.1. Excited synchronous machines

If we redefine the problem of calculating efficiency mappings from "real data" (Table 2.8), for excited synchronous machines, the first two steps would be to first define the operating conditions for the MTPA law and second to deduce the value of the base speed. From there, we would have all the data necessary to calculate the reduced values of the different parameters and quantities. It is in these first few steps that the differences between excited and variable-reluctance synchronous machines become apparent. Table 2.8 defines these first two steps for excited synchronous machines.

Note that the equations that need to be solved to determine the values of ψ_{Opt} (the angle ψ maximizing the torque at the base speed) and Ω_b (the base speed) (Table 2.8) are derived from equations [2.4] and [2.6], which are independent. While using real (non-reduced) values, these equations must be solved simultaneously to determine the values of ψ_{Opt} and Ω_b. They become completely decoupled when defined via the reduced quantities (equations [2.93] and [2.94]). To solve these equations, we assume that the maximum torque is obtained for the maximum armature current magnitude I_{max} and the maximum excitation flux $\Phi_{exc\ max}$. In the case of real quantities, the problem consists of determining the angle ψ_{Opt} and speed Ω_b. Its transposition within the framework of normalized quantities amounts to determining ψ_{Opt} and $V_{n\ max}$.

In the "torque/speed" plane, the computer code developed to plot the efficiency mappings for excited synchronous machines is based on reduced quantities. In executing the code, the angle ψ_{Opt} is determined first before calculating the value of $V_{n\ max}$.

Initial real data	$p, \rho, L_d, R_s, R_f, \Phi_{exc\ max}, V_{max}, I_{max}$
Unknown real quantities	We have two unknowns: ψ_{Opt} (angle ψ maximizing the torque at the base speed) and Ω_b (base speed). We need two equations to determine them: $\begin{cases} \max(\Gamma_{em}) & \text{(equation [2.6])} \\ V = V_{max} & \text{(equation [2.4])} \end{cases}$ ∞ (MPTA control law)
Known reduced values from initial real data	ρ, L_{dn}
Unknown reduced values	$\psi_{Opt}, V_{n\ max}, R_{sn}, R_{fn}$ Note that the normalized values $V_{n\ max}, R_{sn}$ and R_{fn} are all related to the base speed value Ω_b (equations [2.30] and [2.34]). They are therefore known if the base speed is known.

Table 2.8. *Defining the normalized quantities from real ones for excited synchronous machines*

This assumes that the values of R_{sn} and R_{fn} are known, which is also assumed in the script proposed in Figure 2.44. In this script, it should be noted that we have not assumed that the maximum torque is obtained from the maximum armature current amplitude. This amplitude and the angle ψ_{Opt} are determined numerically, and we have verified the hypothesis ($I_n = 1 \Leftrightarrow I = I_{max}$) for various parameter sets.

2.5.4.1.2. Variable-reluctance synchronous machines

Table 2.9 defines the first two steps to go from the "real data" to the reduced data for variable-reluctance synchronous machines. From the normalization system used with variable-reluctance synchronous machines (section 2.4.1.3), the value of $V_{n\,max}$ is known to us ($V_{n\,max} = 1$).

For these machines, it is possible to show that the maximum torque is obtained for the maximum armature current amplitude ($I = I_{max} \Leftrightarrow I_n = 1$). Moreover, it is possible to analytically determine an expression for the angle ψ_{Opt}. As a function of the normalized quantities, this is given by a function of the normalized quantities, by:

$$\psi_{Opt} = \frac{\pi}{2} + \frac{1}{2} \cdot \arcsin\left(\frac{\left(\rho \cdot (\Omega_n \cdot L_{dn})^2 - R_{fn}^2\right)}{\sqrt{\left(\Omega_n \cdot L_{dn} \cdot (\rho+1)\right)^2 \cdot R_{fn}^2 + \left(\rho \cdot (\Omega_n \cdot L_{dn})^2 - R_{fn}^2\right)^2}} \right) \quad [2.104]$$

At the base speed, we have:

$$\begin{cases} \Omega_n = 1 \\ V_n = V_{n\,max} = 1 \end{cases}$$

from which we obtain an equation determining the value of L_{dn}.

In the case of real quantities for these machines, the problem consists of determining the angle ψ_{Opt} and the base speed Ω_b. Analogously with excited synchronous machines, transposing it within the framework of normalized quantities amounts to determining both ψ_{Opt} and L_{dn}.

Given its complexity, we will solve the equation for the value of L_{dn} numerically. Note that a solution exists for this equation only if $R_{sn} < 1$ (Table 2.6), and it can be shown that this is the case for these machines:

$$R_{sn} \in [0, 1) \quad [2.105]$$

and that the value of L_{dn} which solves this equation is unique (see the appendix: https://1drv.ms/b/s!AogSAGtYvycUkHIbc3FbsPoSQjJt?e=U9p85C).

Initial real data	p, ρ, L_d, R_s, R_f, V_{max}, I_{max}
Unknown real quantities	We have two unknowns: ψ_{Opt} (angle ψ maximizing the torque for the base speed) and Ω_b (base speed). We need two equations to determine them: $$\begin{cases} \max(\Gamma_{em}) & \text{(equation [2.6])} \\ V = V_{max} & \text{(equation [2.4])} \end{cases}$$ (MPTA control law) For variable-reluctance synchronous machines, Φ_{exc} must be set to zero in equations [2.4] and [2.6]
Known reduced values from initial real data	ρ, R_{sn}, R_{fn}
Unknown reduced values	ψ_{Opt}, L_{dn}

Table 2.9. *Defining the normalized quantities from real ones for variable-reluctance synchronous machines*

Once the reduced value of each parameter is known, calculating the efficiency mappings takes the same approach as that adopted for excited machines with salient poles. However, here the algorithm is more straightforward [2.106] since there is no excitation flux for variable-reluctance synchronous machines.

For variable-reluctance synchronous machines, the expression of the normalized torque is given by:

$$\Gamma_n = (1-\rho) \cdot L_{dn} \cdot i_{0qn} \cdot i_{0dn} \qquad [2.106]$$

From this last equation [2.106] and equation [2.92] with $k_f = 0$, we come to the following equation:

$$A_{\Gamma 1} \cdot I_n^2 + C_{\Gamma 1} = 0 \qquad [2.107]$$

where

$$\begin{cases} A_{\Gamma 1} = L_{dn} \cdot (1-\rho) \cdot \left(\left(\Omega_n \cdot \rho \cdot L_{dn}/R_{fn} \right) \cdot \cos\psi - \sin\psi \right) \cdot \left(\left(\Omega_n \cdot L_{dn}/R_{fn} \right) \cdot \sin\psi + \cos\psi \right) \\ C_{\Gamma 1} = -\Gamma_n \cdot \left(1 + \rho \cdot \left(\Omega_n \cdot L_{dn}/R_{fn} \right)^2 \right)^2 \end{cases}$$

Note that the value of $C_{\Gamma 1}$ is necessarily negative.

Solving equation [2.107] lets us determine the normalized armature current magnitude. The computer code developed to plot the efficiency mappings in the "torque/speed" plane contains three loops. We have two external loops which are interchangeable, involving the speed and the torque, with only the internal loop involving the angle ψ. For a given value of the angle ψ, the normalized armature current amplitude is given by:

$$I_n = \sqrt{\frac{-C_{\Gamma 1}}{A_{\Gamma 1}}} \qquad [2.108]$$

Figure 2.47 shows the script implemented in order to select the pairs (I_n, ψ), which maximizes the efficiency, while also satisfying the current and voltage constraints for the inner loop. As before, if, for a given operating point, it is not possible to find such a pair (I_n, ψ) satisfying the various constraints, then the efficiency is set to zero $\eta(\Gamma_n, \Omega_n) = 0$.

Once the range of the angle ψ has been completely determined, the internal loop allows us to select the optimal torque (I_n, ψ), which maximizes the efficiency for a given pair (Ω_n, Γ_n).

It should be noted that the efficiency maps can also be calculated in the "power/speed" plane, in a similar method to that in the "torque/speed" plane. An external loop concerning the normalized power $P_n \in [0, 1]$ replaces the torque loop. The "Matlab" scripts developed to compute the efficiency maps for each type of synchronous machine are provided in the appendix (https://1drv.ms/b/s!AogSAGtYvycUkHIbc3FbsPoSQjJt?e=U9p85C), and we will study their validity in the next section. These scripts allow for the efficiency maps to be plotted in the respective "torque/speed" and "power/speed" planes.

2.5.5. *Validity of the tools developed and the contribution towards hybrid excitation*

In this section, the validity of the code developed to plot the efficiency maps is studied, first by comparing it to the previous lossless models, and then by comparing it to the models developed for non-salient and salient pole machines. Recall that machines with non-salient poles are special cases of those with salient poles.

Once we have checked its validity, the computer code can be used to present the contribution that hybrid excitation provides (Amara 2001; Amara et al. 2009, 2019, 2021a).

Start

If A_Π is negative $\Rightarrow \eta(\Gamma_n, \Omega_n, \psi, I_n) = 0$

Otherwise

If $I_n > 1 \Rightarrow \eta(\Gamma_n, \Omega_n, \psi, I_n) = 0$

Otherwise

Knowing ψ and $I_n < 1$, it is possible to calculate V_n and the efficiency

If $V_n > V_{n\,max} \Rightarrow \eta(\Gamma_n, \Omega_n, k_f, \psi) = 0$

Otherwise $\eta = \eta(\Gamma_n, \Omega_n, k_f, \psi, I_n)$

End

Figure 2.47. *Script used to select the combination (I_n, ψ) (inner loop)*

2.5.5.1. *Validity of the computer code*

The validity of the developed numerical tools is checked in two steps: first, the operating envelopes in the "torque/speed" plane obtained from the lossless models are compared to those obtained from the numerical tools, developed specifically for calculating efficiency maps. The comparison of these envelopes for hybrid excited synchronous machines with non-salient poles, and with and without excitation flux control, is done first. The programs that were developed specifically to calculate the efficiency maps for non-salient pole machines (Numerical Approach 1) will be compared to those developed for salient pole machines (Numerical Approach 2), given that non-salient pole machines are special cases of salient pole ones, in addition to the optimal control strategies determined via the lossless models. Losses will be neglected in the tools developed to calculate the efficiency maps: $R_{en} = R_{sn} = 0$, and $R_{fn} \to +\infty$. Also, in the first step, the envelopes are compared in the case of salient pole machines. Finally, in the second step, efficiency maps from the two programs developed ("Numerical Approach 1" and "Numerical Approach 2") for non-salient pole machines are compared.

Figure 2.48 compares the envelopes for a non-salient pole permanent magnet synchronous machine with $L_{dn} = 0.5$ in the "torque/speed" and "power/speed" planes. These envelopes are obtained from the lossless model and the two numerical approaches. Figure 2.49 presents a similar comparison for a permanent magnet machine with $\rho = 1$ and $L_{dn} = 1.5$. The three approaches provide similar results, allowing us to ascertain the relevancy of each numerical approach presented and for computer script developed.

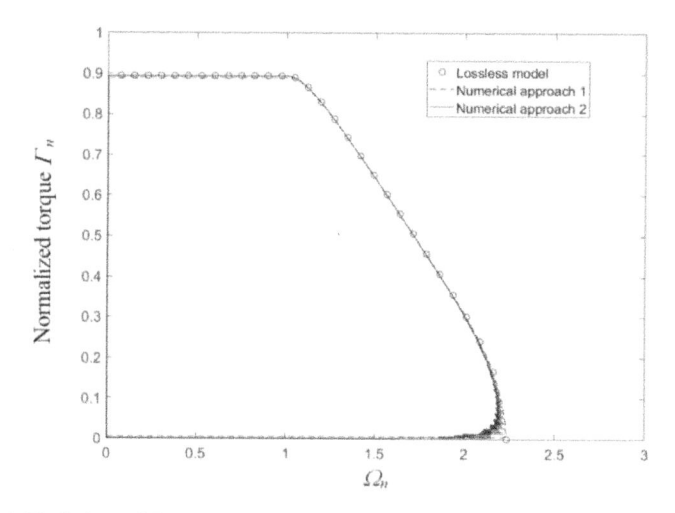

a) Variation of the normalized torque as a function of the normalized speed

b) Variation of the normalized power as a function of the normalized speed

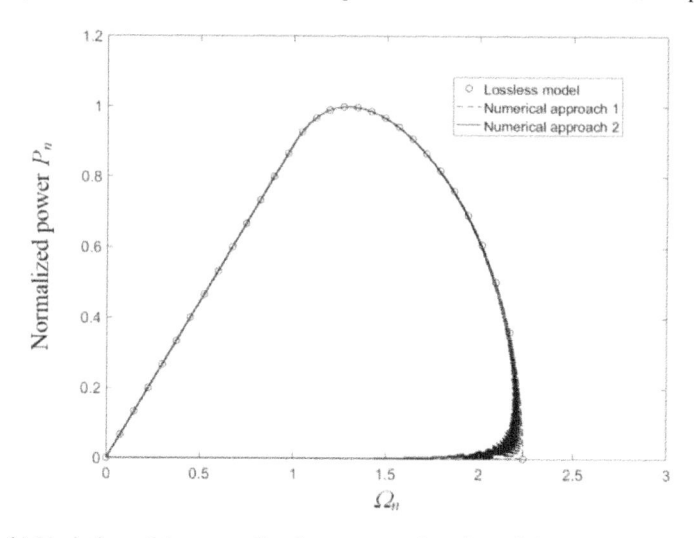

Figure 2.48. *Variations of the maximum normalized torque (a), and of the maximum normalized power (b), with respect to the normalized speed for a machine with ($\rho = 1$, $L_{dn} = 0.5$)*

a) Variation of the normalized torque as a function of the normalized speed

b) Variation of the normalized power as a function of the normalized speed.

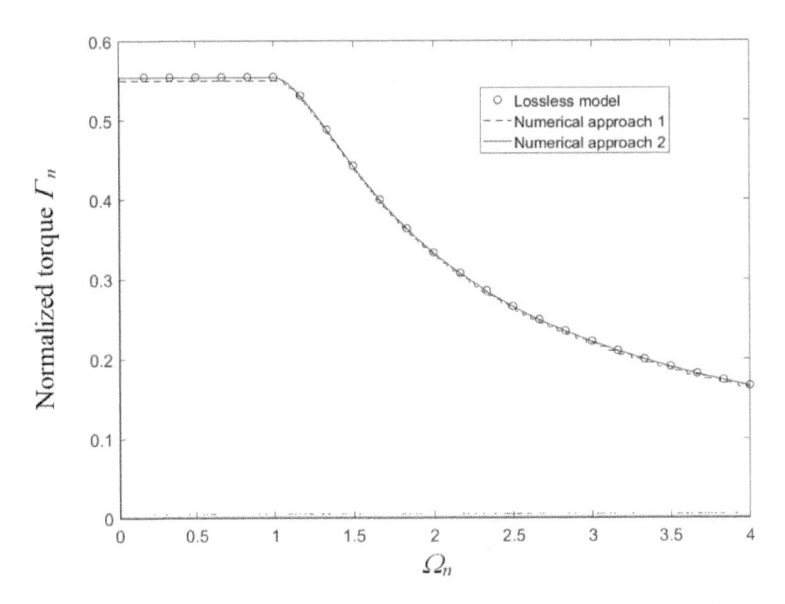

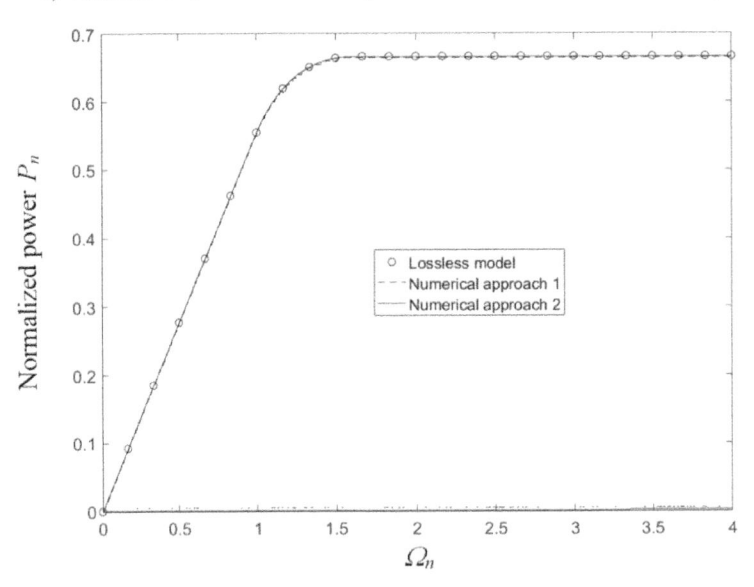

Figure 2.49. *Variations of the maximum normalized torque (a), and of the maximum normalized power (b), with respect to the normalized speed for a machine with ($\rho = 1$, $L_{dn} = 1.5$)*

The numerical noise, which is clearly visible in Figure 2.48, comes from the curves which result from "Numerical Approach 2". This noise is linked to the discretization intervals of (Ω_n, Γ_n, ψ) that are covered by the different loops (three loops), which we have voluntarily chosen to be less dense in order to reduce the execution time. It should be recalled that "Numerical Approach 1" only requires two discretization intervals (Ω_n, Γ_n), and thus two loops.

Figure 2.50 compares the envelopes in the "torque/speed" and "power/speed" planes for a hybrid excited synchronous machine with smooth poles and $L_{dn} = 0.5$. The curves that arise from the three approaches coincide perfectly. We see that the hybrid excitation lets us extend the operation to higher speeds, unlike with permanent magnet machines (Figure 2.48). Hybrid excitation, or to be precise, excitation flux control, does not allow for the envelopes to be extended in the case of a machine with $L_{dn} = 1.5$.

Next, we will investigate the validity in the case of synchronous machines with salient poles. We will repeat the cases studied in section 2.4 with the lossless model, in order to compare them with "Numerical Approach 2".

Figures 2.51 to 2.56 compare the curves shown in Figures 2.22, 2.25, 2.30, 2.33, 2.35 and 2.39, respectively, to those obtained by using "Numerical Approach 2".

The numerical noise is clearly notable again in the curves obtained via "Numerical Approach 2", especially in Figures 2.51, 2.52 and 2.55. Overall, this approach yields similar results to those obtained from the lossless model, which was based on a formally analytical approach.

For the second step of our validity study, the efficiency maps obtained from "Numerical Approach 1" and "Numerical Approach 2" are compared for three non-salient poles machines ($\rho = 1$): a permanent magnet machine (Figure 2.57), a machine with wound excitation (Figure 2.58) and finally a machine with hybrid excitation (Figure 2.59). These machines share the same following parameters: $L_{dn} = 0.5$, $R_{sn} = 0.1$ and $R_{fn} = 20$. The wound and hybrid excited machines also share the following parameters: $k_{en} = 1$, $R_{en} = 1$ and $\beta = 27$. The hybridization ratio α is equal to 1 for the hybrid excited machine and is 0 ($\alpha = 0$) for the wound excitation machine. These parameters were taken from a hybrid excited prototype machine (Amara 2001; Amara et al. 2004b).

Again, we see that the two numerical approaches give similar results.

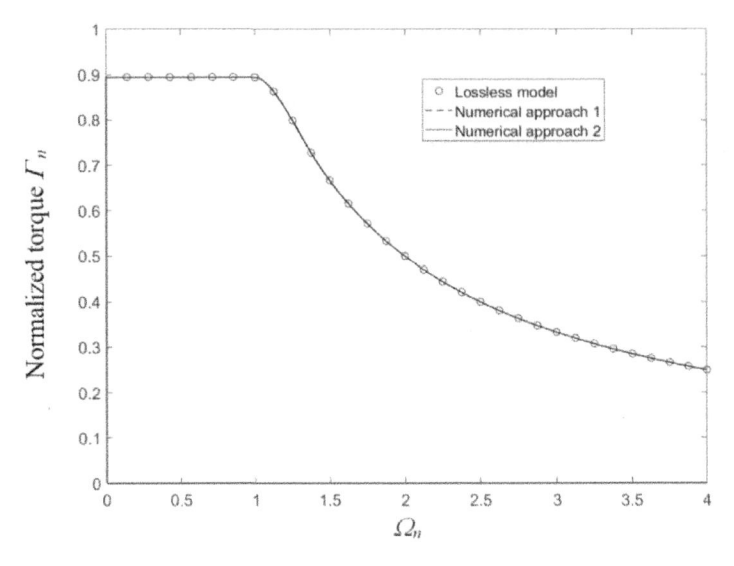

a) Variation of the normalized torque as a function of the normalized speed

b) Variation of the normalized power as a function of the normalized speed

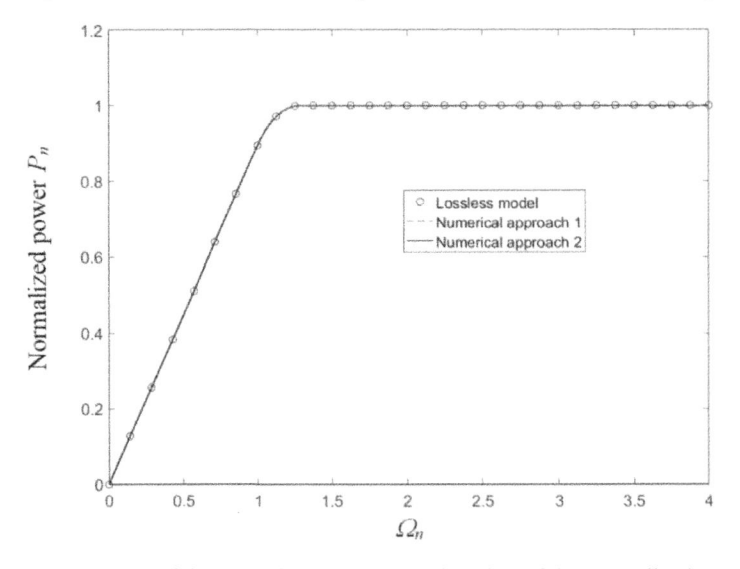

Figure 2.50. *Variations of the maximum normalized torque (a), and of the maximum normalized power (b), with respect to the normalized speed for a machine with ($\rho = 1$, $L_{dn} = 0.5$)*

a) Variation of the normalized torque as a function of the normalized speed

b) Variation of the normalized power as a function of the normalized speed

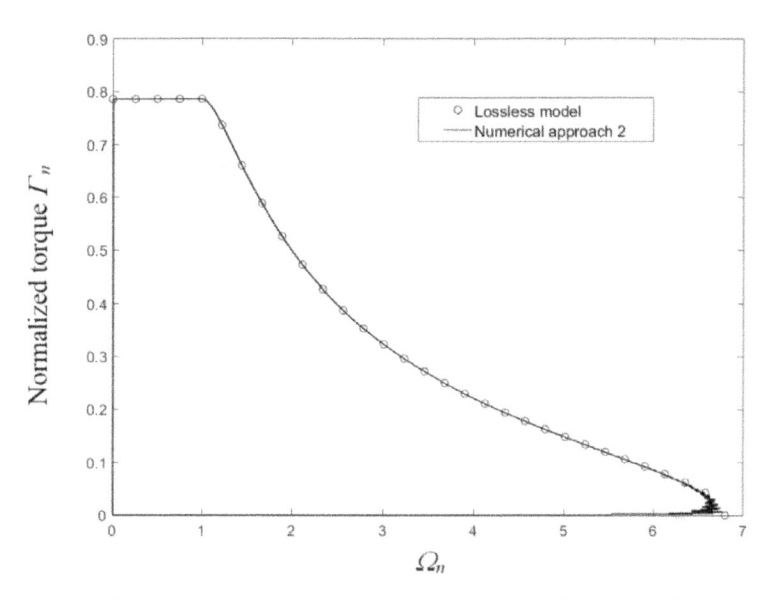

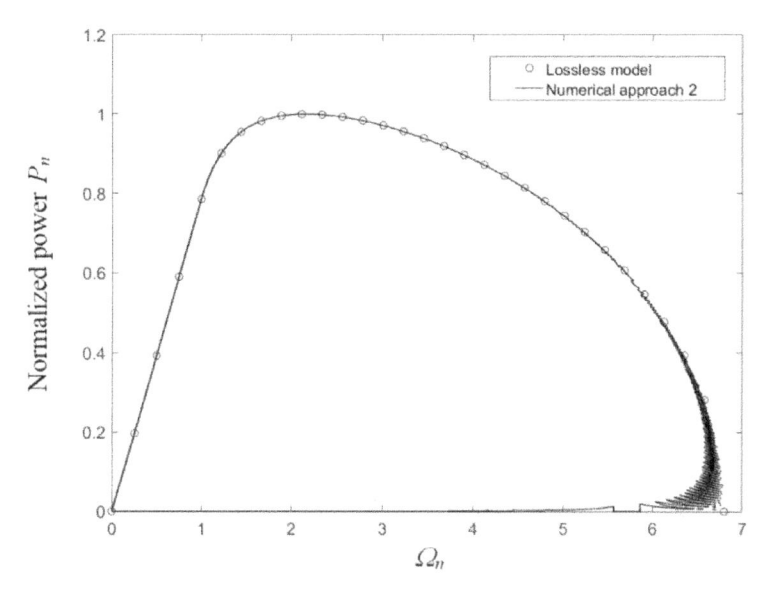

Figure 2.51. *Variations of maximum normalized torque (a), and of the maximum normalized power (b), with respect to the normalized speed for a machine with ($\rho = 1.5$, $L_{dn} = 0.8$)*

a) Variation of the normalized torque as a function of the normalized speed

b) Variation of the normalized power as a function of the normalized speed

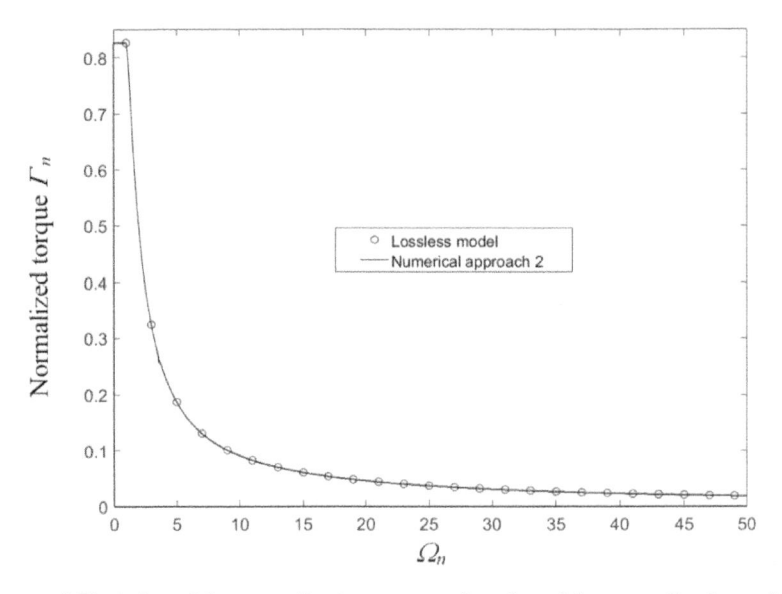

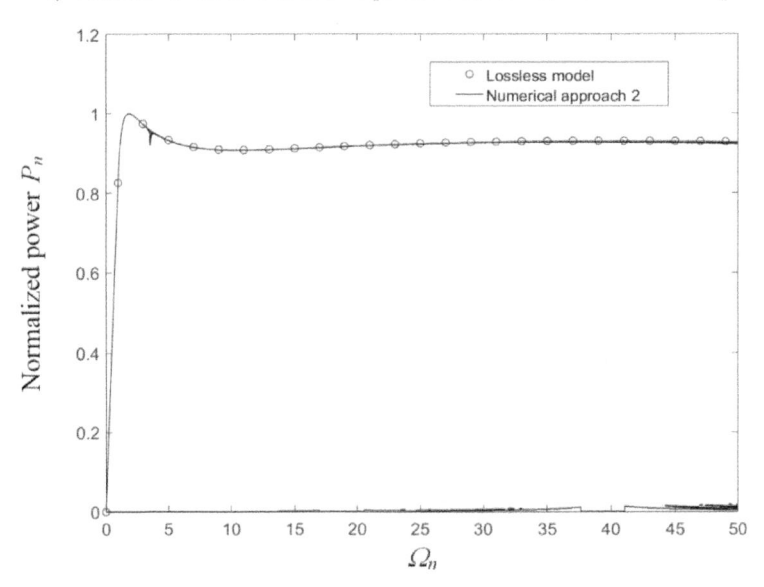

Figure 2.52. *Variations of the maximum normalized torque (a), and of the maximum normalized power (b), with respect to the normalized speed for a machine with ($\rho = 0.2$, $L_{dn} = 0.994$)*

a) Variation of the normalized torque as a function of the normalized speed

b) Variation of the normalized power as a function of the normalized speed

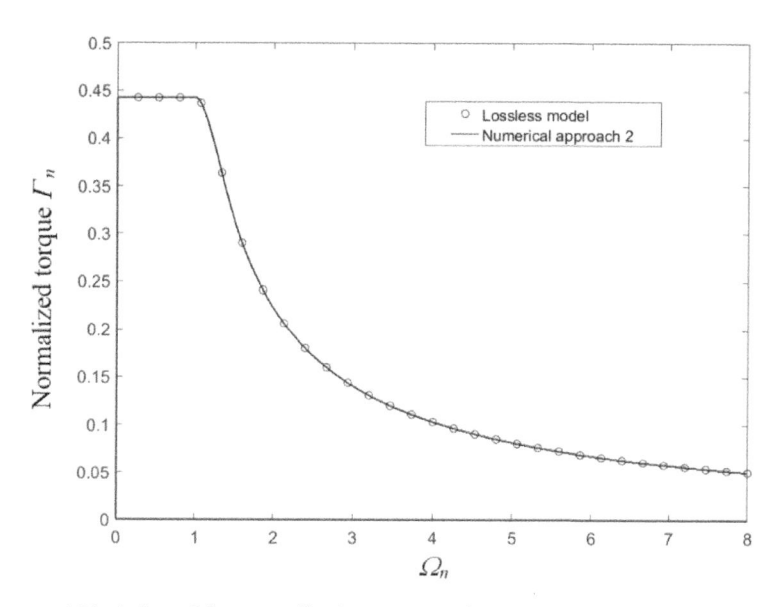

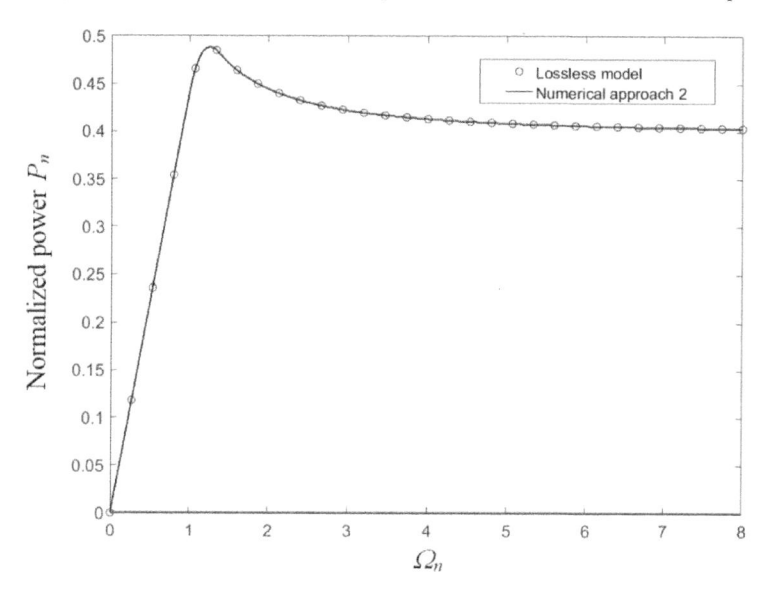

Figure 2.53. *Variations of the maximum normalized torque (a), and of the maximum normalized power (b), with respect to the normalized speed for a machine with ($\rho = 1.5$, $L_{dn} = 2.5$)*

a) Variation of the normalized torque as a function of the normalized speed

b) Variation of the normalized power as a function of the normalized speed

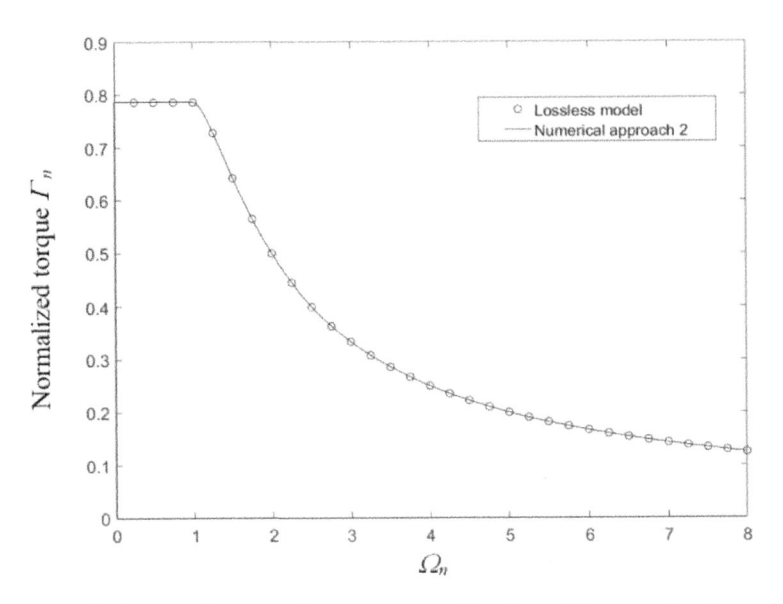

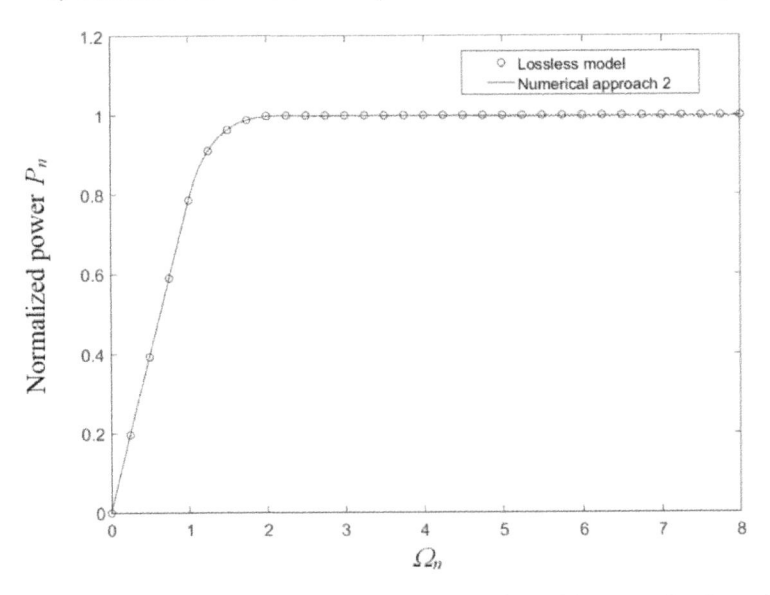

Figure 2.54. *Variations of the maximum normalized torque (a), and of the maximum normalized power (b), with respect to the normalized speed for a machine with (ρ = 1.5, L_{dn} = 0.8)*

a) Variation of the normalized torque as a function of the normalized speed

b) Variation of the normalized power as a function of the normalized speed

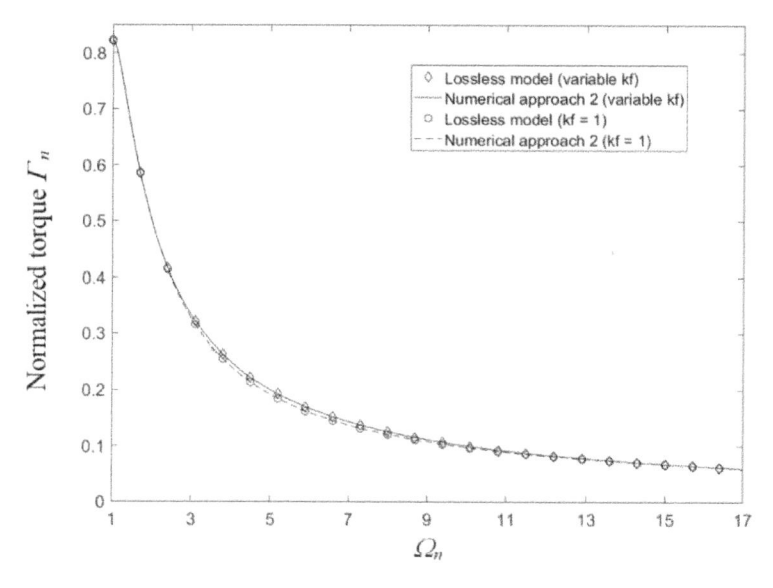

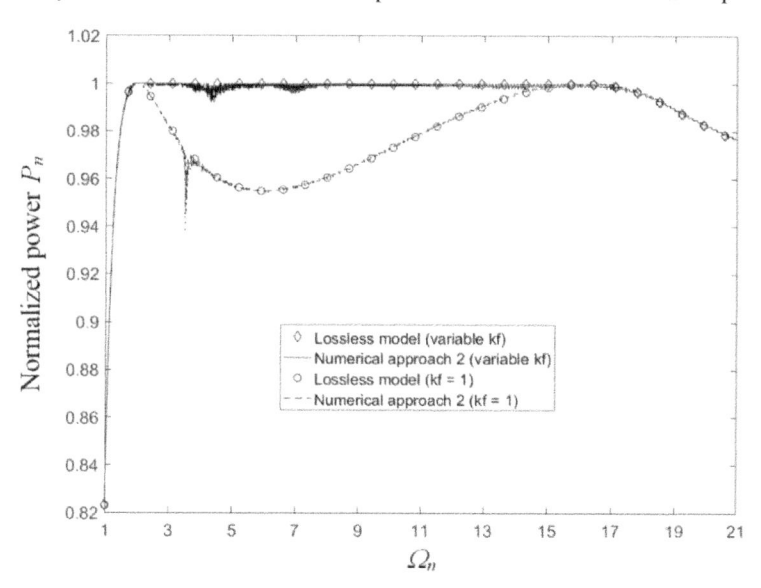

Figure 2.55. *Variations of the maximum normalized torque (a), and of the maximum normalized power (b), with respect to the normalized speed for a machine with (L_{dn} = 1.1, ρ = 0.15)*

a) Variation of the normalized torque as a function of the normalized speed

b) Variation of the normalized power as a function of the normalized speed

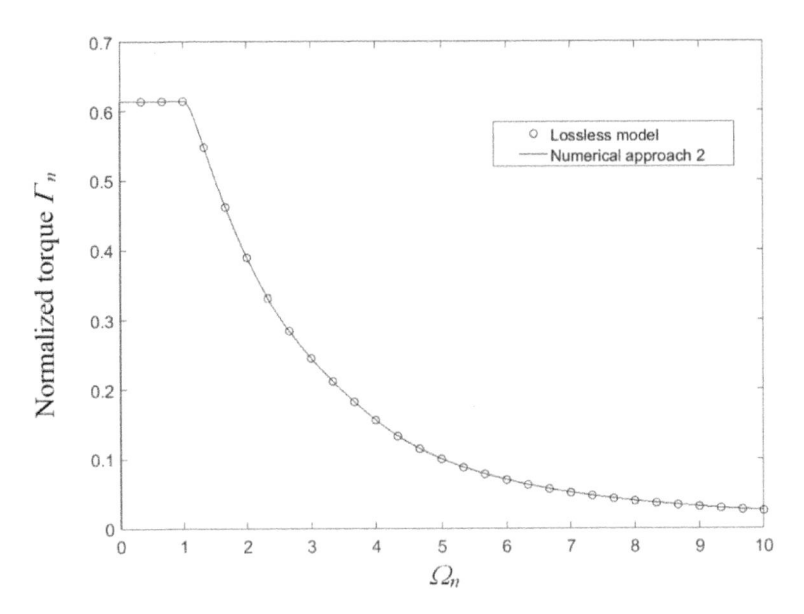

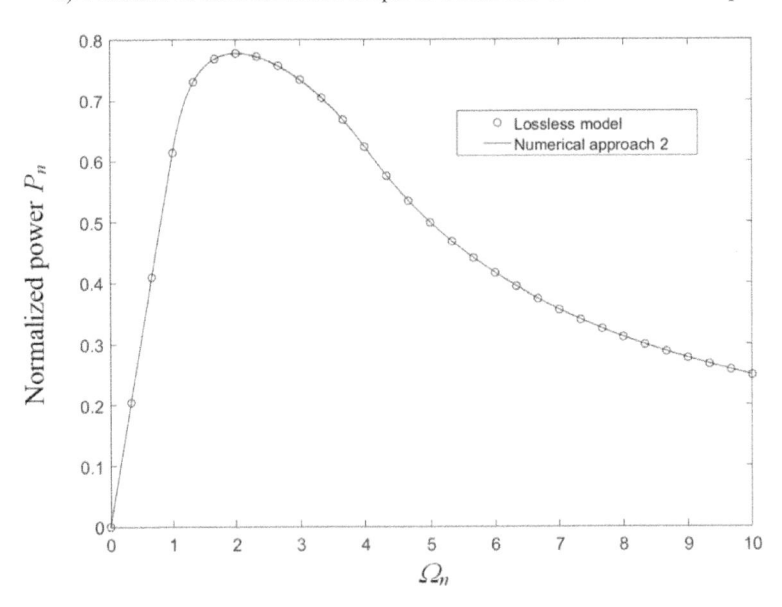

Figure 2.56. *Variations of the maximum normalized torque (a), and of the maximum normalized power (b), with respect to the normalized speed for a machine with ($p = 8$)*

a) Numerical approach 1

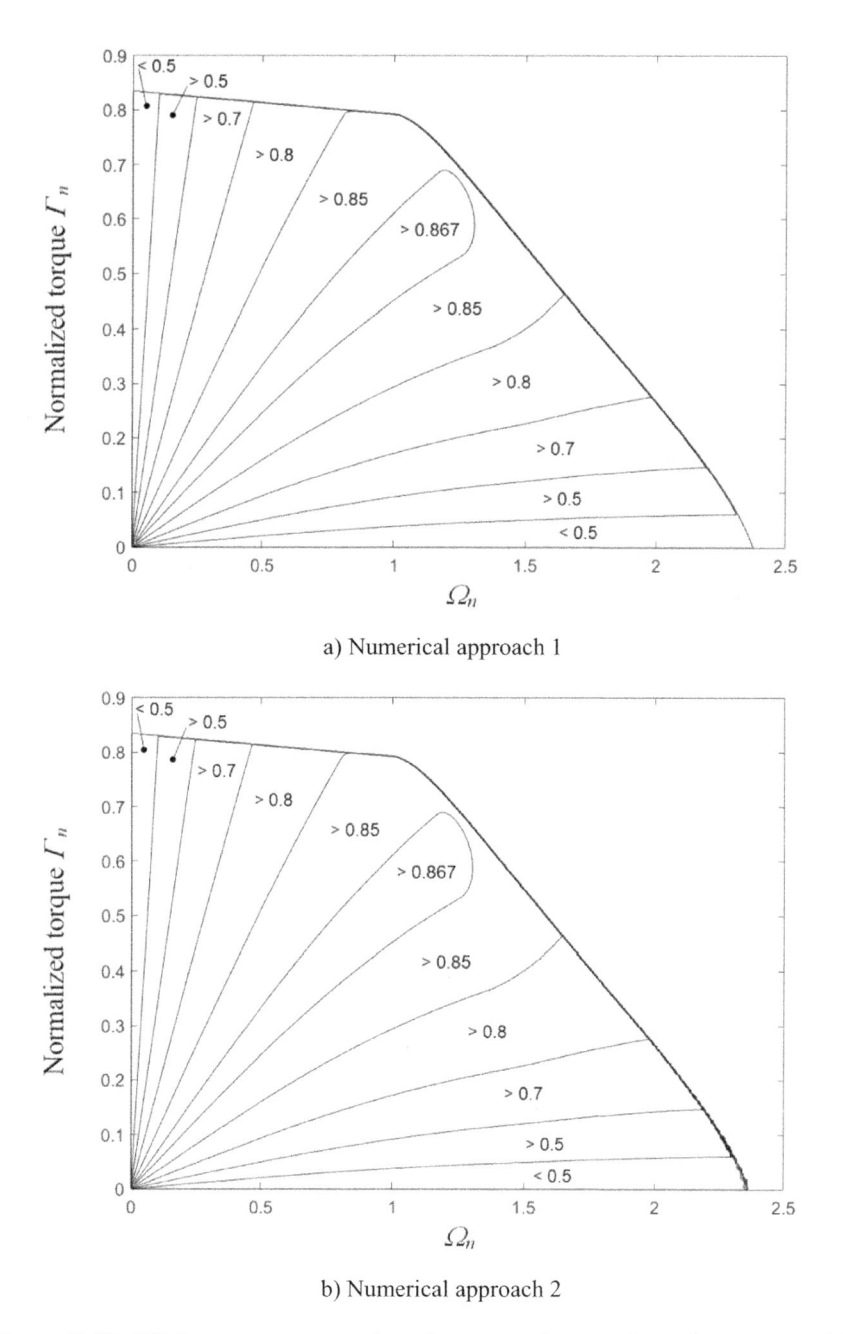

b) Numerical approach 2

Figure 2.57. *Efficiency map comparison (permanent magnet synchronous machines)*

a) Numerical approach 1

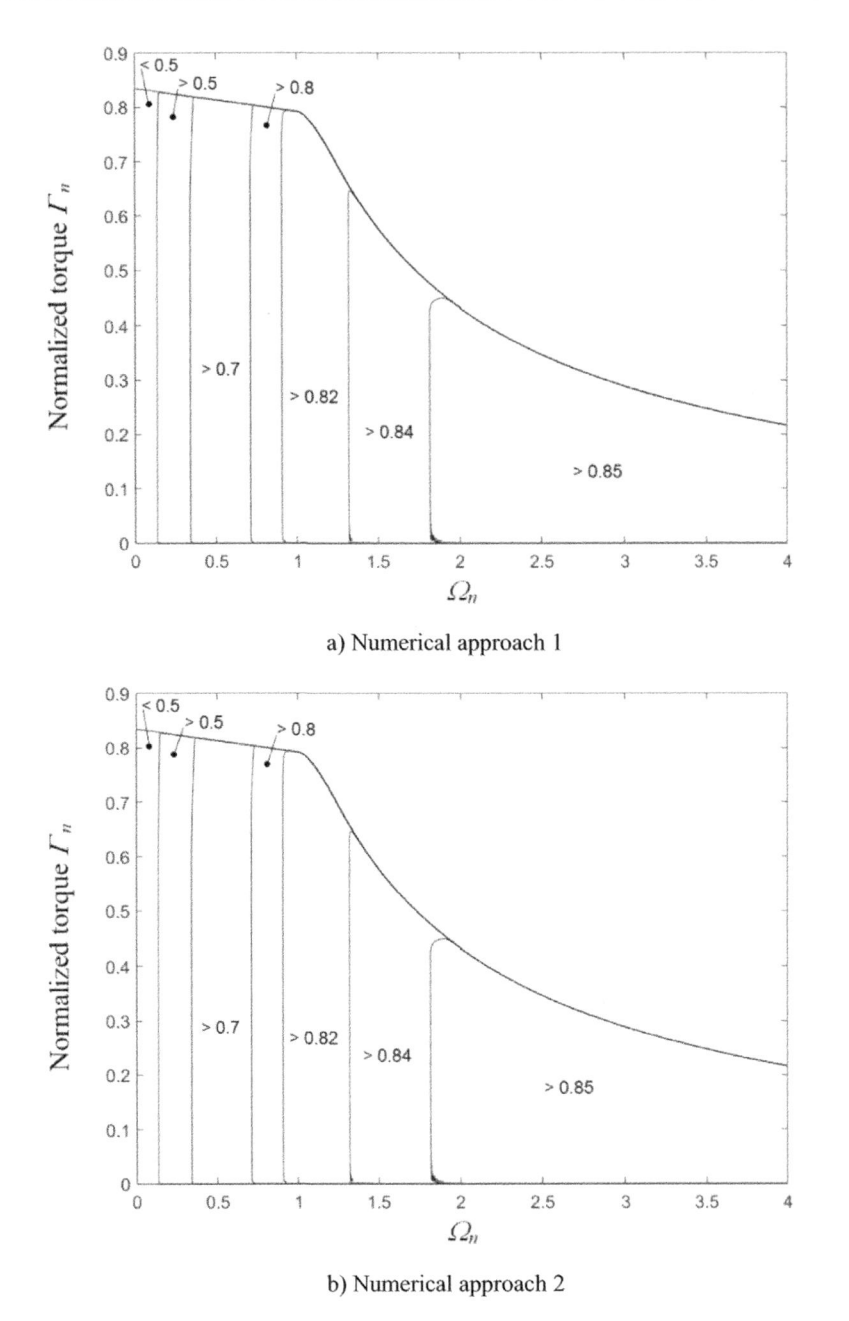

b) Numerical approach 2

Figure 2.58. *Efficiency map comparison (wound excitation synchronous machines)*

a) Numerical approach 1

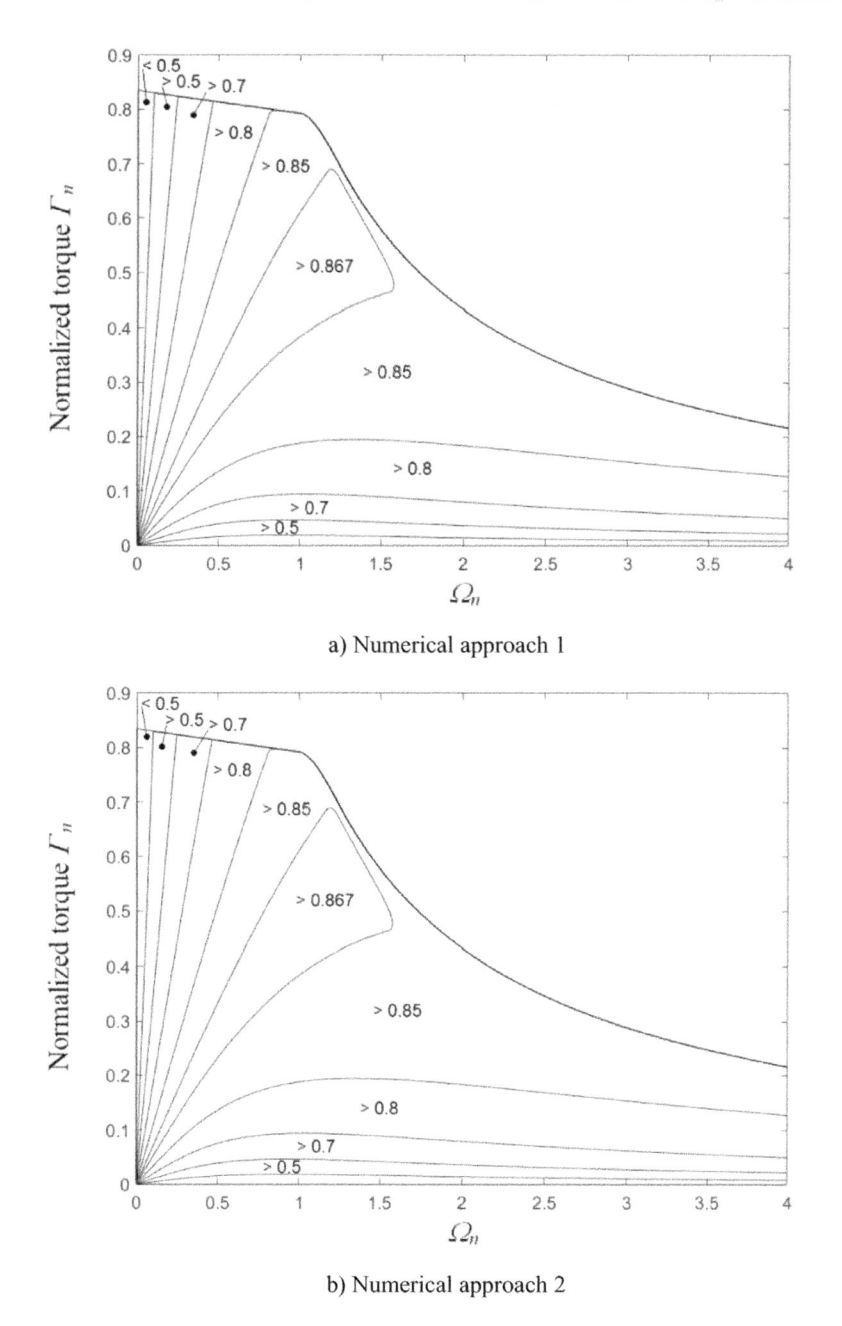

b) Numerical approach 2

Figure 2.59. *Efficiency map comparison (hybrid excited synchronous machines)*

2.5.5.2. *The contribution coming from hybrid excitation*

In addition to improving high-speed operations, hybrid excitation also allows for the energy efficiency of synchronous machines to be improved by choosing the optimal hybridization ratio α (Amara et al. 2009, 2019, 2021a). An increase in energy efficiency allows for an increase in the range of a vehicle (Biais and Langry 1998).

We have already demonstrated the contribution that hybrid excitation provides in several publications (Amara et al. 2009, 2019, 2021a). Here, we will recall the most important results illustrating this. Figure 2.60 compares the efficiency maps for two hybrid excited machines whose parameters are given in Table 2.10. It is important to note that the variation in the hybridization ratio α modifies how the efficiency zones are distributed. It is therefore important to optimize the hybridization ratio α, in order to maximize the efficiency in the most frequently used areas.

Let us consider a hybrid excited machine which shares the same parameters as the two previous machines, but for which we wish to optimize the hybridization ratio value α so that the torque efficiency ($\Omega_n = 2$, $\Gamma_n = 0.2$) is maximized.

Parameters	Machine 1	Machine 2
p	1	
L_{dn}	0.5	
R_{sn}	0.1	
R_{fn}	20	
R_{en}	1	
k_{en}	1	
β_0	27	
α	1	0.8

Table 2.10. *Parameters of the hybrid excited synchronous machines*

By executing the algorithm presented in Figure 2.43, the curves in Figure 2.61 are obtained. The efficiency of the pair ($\Omega_n = 2$, $\Gamma_n = 0.2$) is maximized for a hybridization ratio value of $\alpha = 0.5041$. Figure 2.62 presents the efficiency maps for this hybridization ratio, and the improvement in the efficiency for the region around the operating point ($\Omega_n = 2$, $\Gamma_n = 0.2$) can be clearly seen.

a) $\alpha = 1$ (Machine 1)

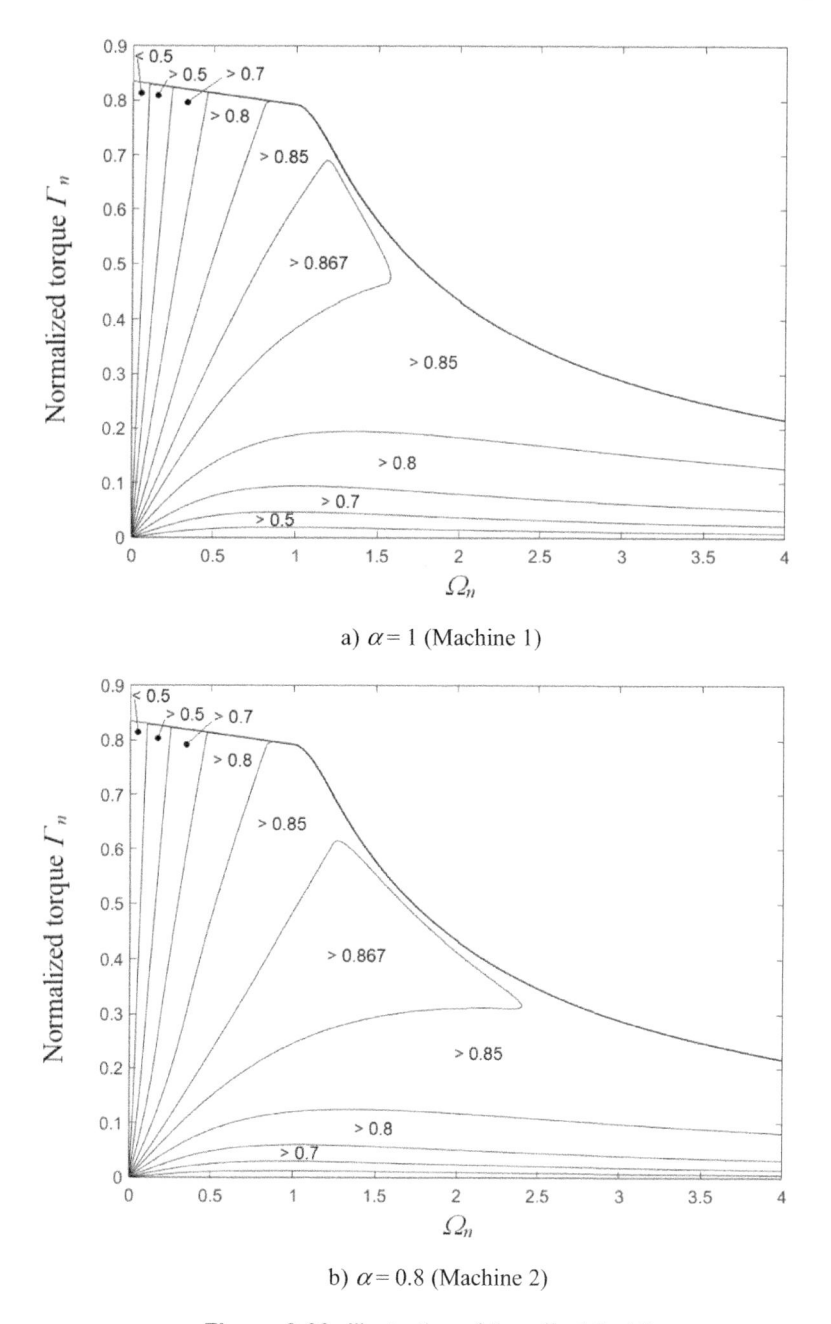

b) $\alpha = 0.8$ (Machine 2)

Figure 2.60. *Illustration of the effect that the hybridization ratio has on the efficiency maps*

2.6. Conclusions and perspectives

After the first chapter was dedicated to hybrid excited structures, where we saw how the principle of hybrid excitation was incorporated into the design of synchronous structures, in this second chapter, we have now seen how hybrid excitation allows for the operation on the entire "torque/speed" or "power/speed" plane to be optimized. The first chapter focused on the hybrid excited synchronous machine component, whereas this second one focused on the aspects of the system, by considering the machine in its environment, in particular when associated with a power electronics converter.

The formal or numerical mathematical developments presented in this chapter can be used for all types of synchronous machines since the permanent magnet excited synchronous machines models, the wound excitation synchronous machines and the variable-reluctance synchronous machines models are all derived from the same general hybrid excited machine model. The computer code that allows the reader to reproduce all of the curves presented in the figures of this chapter themselves is available to download in the appendix (https://1drv.ms/b/s!AogSAGtYvycUkHIbc3Fbs PoSQjJt?e=U9p85C).

Magnetic saturation is an important aspect of the behavior of electrical machines, and it has not been explicitly mentioned in the previous sections. However, since it essentially induces a variation in the inductance values, and the parametric study carried out covers the whole (L_{dn}, ρ) plane, its effects have been mentioned explicitly. Thus, knowing the effects of saturation, it is relatively simple to predict their impact on performance.

In this chapter, we have detailed an approach used to study hybrid excited synchronous machines. It consisted of developing a set of tools to study these machines and then comparing them with other families of synchronous machines. It has been demonstrated that the hybrid excitation principle allows for the dimensioning of the drives that are integrated within these machines to be optimal. This principle visibly improves the power factor and the dimensioning of the power converter–machine assembly, in addition to the energy efficiency.

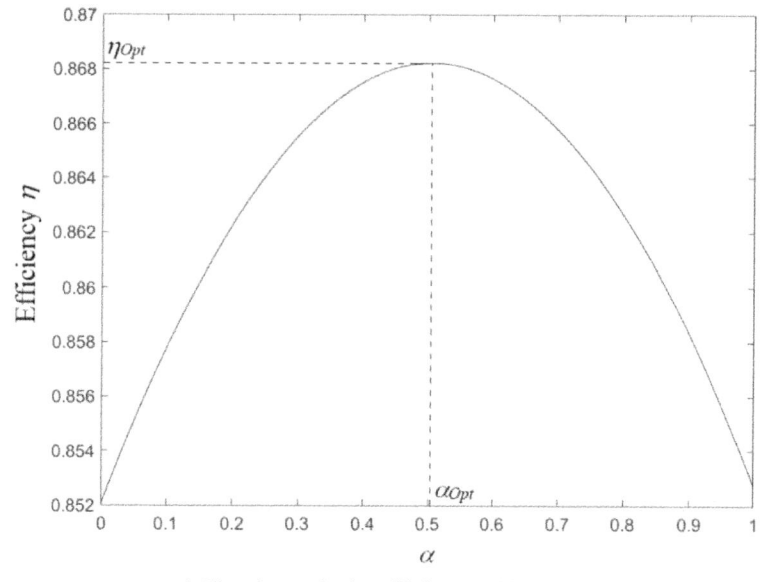

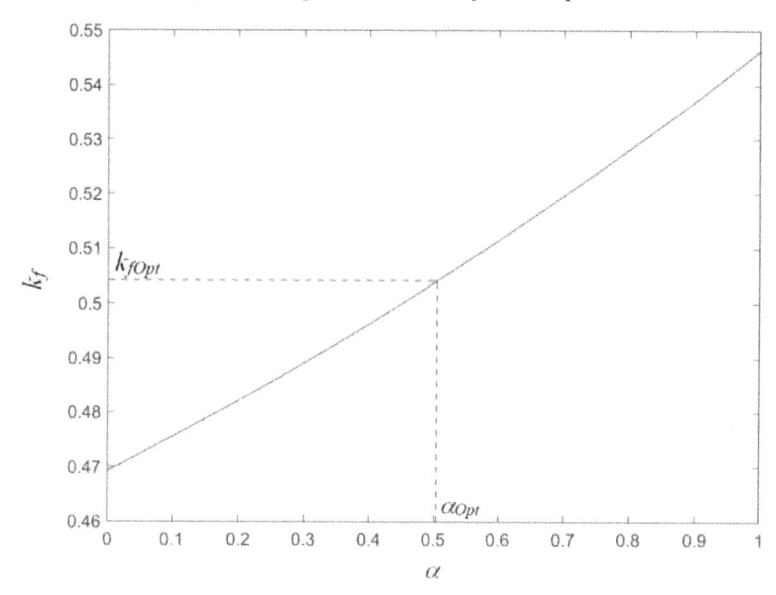

a) The change in the efficiency with respect to α

b) Variation of k_f with respect to α

Figure 2.61. *Optimization of the hybridization ratio which optimizes the efficiency point ($\Omega_n = 2$, $\Gamma_n = 0.2$)*

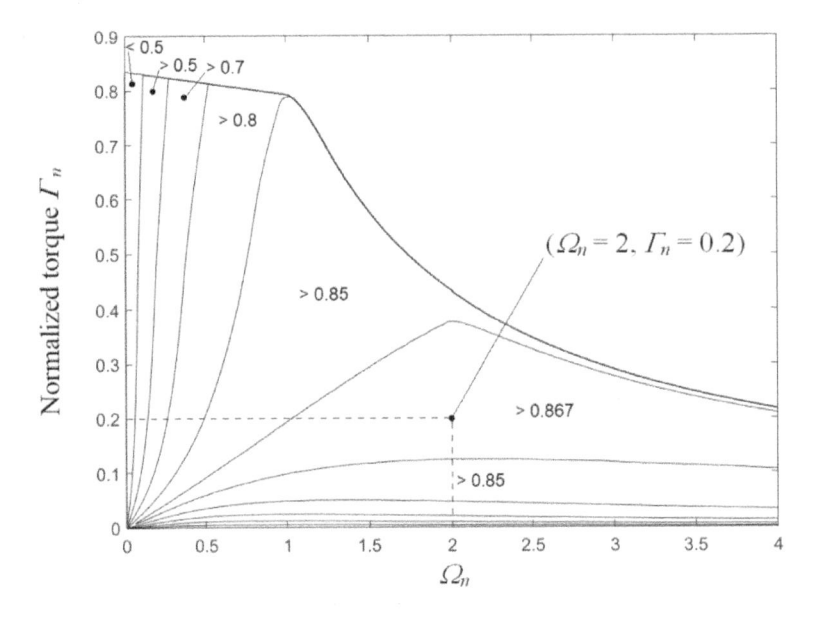

Figure 2.62. Efficiency maps for a machine with $\alpha = \alpha_{Opt} = 0.5041$

It would now be interesting to repeat these developments in the case of excited synchronous machines with shifted inductance axes (Figure 2.63), which allows for the torque density to be improved (Zhao et al. 2015a, 2015b; Yang et al. 2017), but whose model is slightly different from a conventional synchronous machine. Indeed, compared to the classical excited synchronous machine where the maximum reluctance torque does not coincide with the maximum interaction torque between the excitation flux and the armature current, in these new structures, the two maxima do coincide, thus allowing for the torque to increase (Zhao et al. 2015a, 2015b; Yang et al. 2017).

In the models, it would also be interesting to include a more detailed consideration of electromagnetic losses, in particular iron losses, and to incorporate the mechanical losses into them, which are relatively important for high-speed operations.

Using more precise models would limit the possibility of finding formal analytical solutions. Therefore, numerical approaches are necessary in this case. If we were to consider more precise models, the numerical approach would not increase the complexity by too much, since the number of loops remains constant.

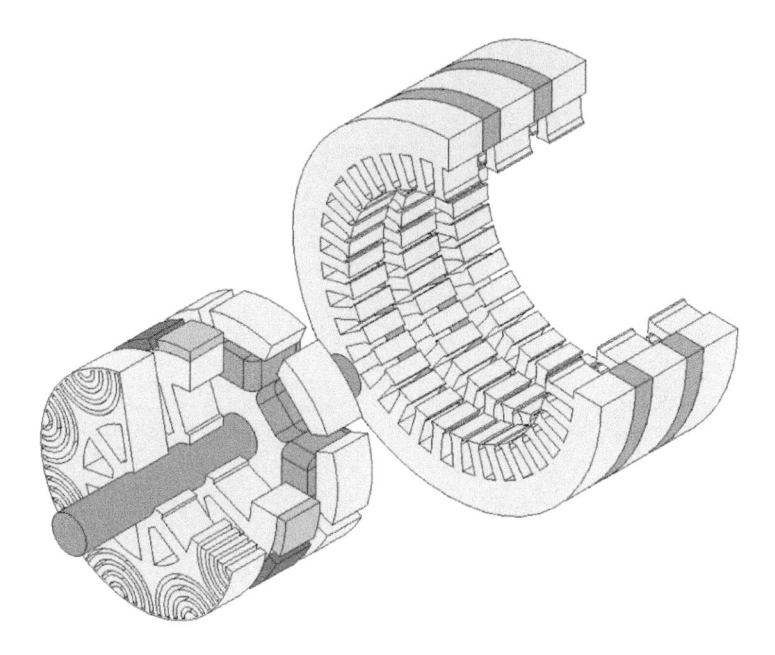

Figure 2.63. *Hybrid excited synchronous machine with offset inductor axes (Amara and Hlioui 2019)*

3

Experimental Studies of Hybrid Excited Synchronous Machines

3.1. Introduction

Chapter 1 is dedicated to the principle of hybrid excitation and how it is integrated into synchronous machines, and Chapter 2 is dedicated to controlling synchronous motors and the contribution that hybrid excitation provides. This third chapter concerns experimental studies of two hybrid excited machine prototypes.

These two machines perfectly illustrate how the concept has matured in our respective laboratories (SATIE and GREAH) since the late 1990s until recently (Amara 2001; Vido 2004; Hlioui 2008; Takorabet 2008; La Barrière 2010; Nedjar 2011; Gaussens 2013; Dupas 2016; Nasr 2017; Diab et al. 2021). The first structure was studied as part of a PhD thesis at LESiR/SATIE (Amara 2001), with applications in transport as a backdrop, and the second was part of a project dedicated to renewable marine energy at GREAH (Diab et al. 2021; http://neptune-project.com/).

These are exclusively rotary machines. The first machine has a 3D structure, with magnetic flux paths in three dimensions, whereas the second has a 2D structure, with fluxes flowing essentially in the planes perpendicular to the axis of rotation. The second machine is a flux-switching machine. Hybrid excited flux-switching machines are particularly popular because of the advantages that they offer, with the presence of all the

---

For a color version of all the figures in this chapter, see www.iste.co.uk/amara/hybrid.zip.

magnetic flux sources on one armature, often the stator, and a second armature that is completely passive, often the moving part.

The motor and generator functions will be discussed. These experimental studies will allow us to validate, as well as illustrate concretely, the characteristics and contributions that hybrid excitation provides.

3.2. Machine 1

This structure was developed at the LESiR/SATIE laboratory (Amara 2001). It is part of a series of structures developed and studied to validate the principle of hybrid excitation and evaluate certain design choices (Amara 2001). The dimensions of these machines were based on standard considerations of magnetic circuits, such as the conservation of magnetic flux. All these structures share the same stator, which is made from standard laminations generally available on the market.

Figure 3.1 presents a view of the structure, which we will call "Machine 1". We can see the permanent magnets and the excitation coils, and the armature windings are not shown. Table 3.1 lists some important features of the structure. More information about this machine will be provided in the following sections.

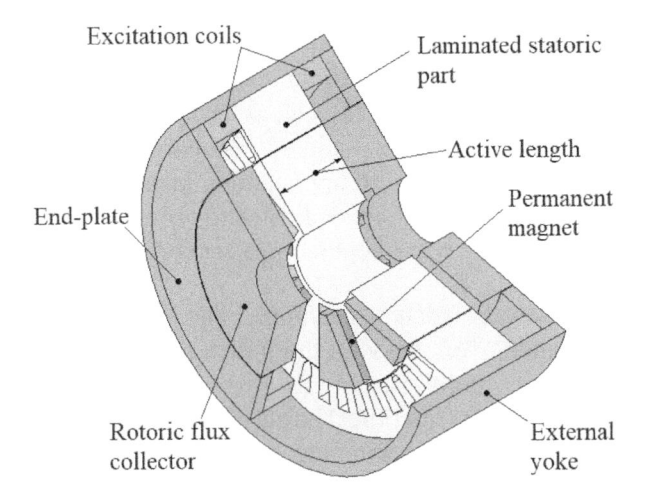

Figure 3.1. *3D cross-sectional view of the first hybrid excited structure studied (Machine 1) (Amara 2001)*

Parameter	Value
External diameter of the stator (mm)	184
Internal diameter of the stator (mm)	115
Air-gap thickness (mm)	0.5
Axial length of the machine (mm)	115
Active length (mm)	40
Number of poles/slots	12/36
Number of phases	3
Number of excitation coil turns for each coil (two coils)	150
Number of armature winding turns	33
Resistance of the wound inductor (Ω)	2.7
Resistance of one armature phase (Ω)	0.76
Type of permanent magnet	Ferrite
Magnet dimensions: thickness, height, length (mm)	6, 40, 44

Table 3.1. *Main parameters of "Machine 1"*

This first structure can be assembled into two different configurations (Vido et al. 2005a, 2005b):

1) hybrid homopolar excitation, where the wound excitation flux acts on only one type of magnetic pole;

2) hybrid bipolar excitation, where the wound excitation flux acts on both types of magnetic pole at the same time.

These two configurations will be compared for no-load operation in section 3.2.3.

3.2.1. *Structure and operating principles*

This machine can be described as a parallel hybrid excited machine with a 3D structure, with the magnetic fluxes flowing in three dimensions. The ferromagnetic circuit of this structure is composed of laminated parts in the stator and rotor, as well as massive parts (the outer yoke and flanges for the stator, and rotor flux collectors for the rotor) offering low-reluctance paths for the wound excitation flux. The massive parts are used in regions of the machine where the magnetic flux changes direction, which would be difficult to achieve with laminations.

We now present the operation of this structure in its bipolar configuration, as it allows for a wider variation range in the excitation flux than the homopolar version (Vido et al. 2005a, 2005b).

Figure 3.2(a) shows the main path of the permanent magnet excitation flux, which participates in the torque production. Figure 3.2(b) shows, on the other hand, a path that can be described as flux leakage since it does not participate in producing torque. This path does not cross the armature windings.

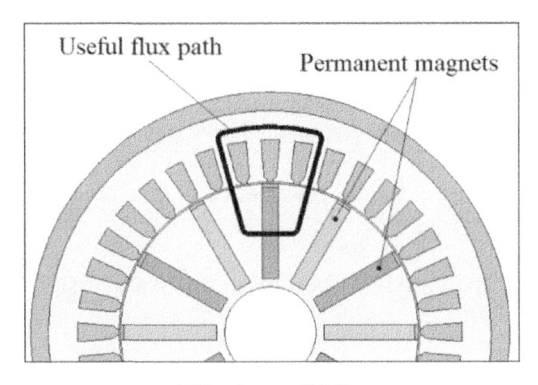

a) Bipolar useful flux.

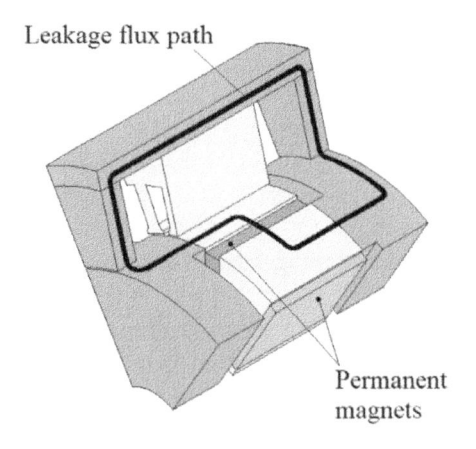

b) Leakage flux.

Figure 3.2. *Flux paths generated by the permanent magnets*

The leakage flux is much lower in the case of the homopolar configuration. In the homopolar configuration (Figure 3.3), the rotor flux collectors are aligned along the machine axis, whereas they are offset by one pole pitch in the bipolar configuration (Figure 3.2(b)). The equivalent path in the homopolar configuration is much more reluctant since the flux lines pass through the air-gap between the laminated rotor pole and the rotoric flux collector (Figure 3.3).

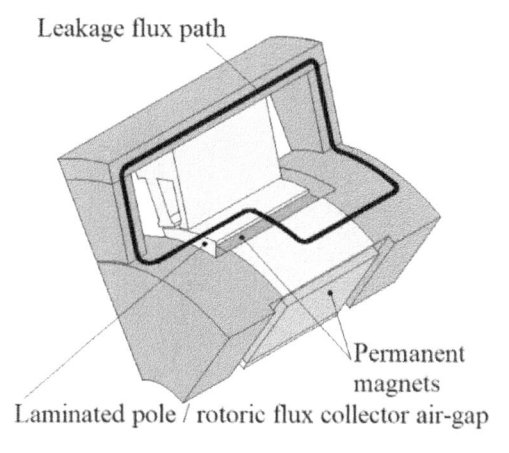

Figure 3.3. *Leakage flux paths in the homopolar configuration*

Figure 3.4 shows (for the bipolar configuration) homopolar flux paths coming from the permanent magnets, so called because they create only one type of magnetic pole in the active part of the machine (Figure 3.1).

Figure 3.5 shows the flux paths coming from the wound excitation. The machine has two annular excitation coils, with each coil acting on one type of magnetic pole. The flux created by an excitation coil passes through the air-gap of the active part only once (homopolar path). Depending on the direction of the excitation current, coils can be used to either increase or decrease the excitation flux through the armature windings.

For the homopolar version of the structure (Figure 3.3), the fluxes coming from the two excitation coils act only on one type of magnetic pole. It is then straightforward to see that the variational range of the excitation flux for the homopolar version is less extended than that for the bipolar configuration. This is essentially due to the magnetic behavior of the machine when the excitation flux is increased.

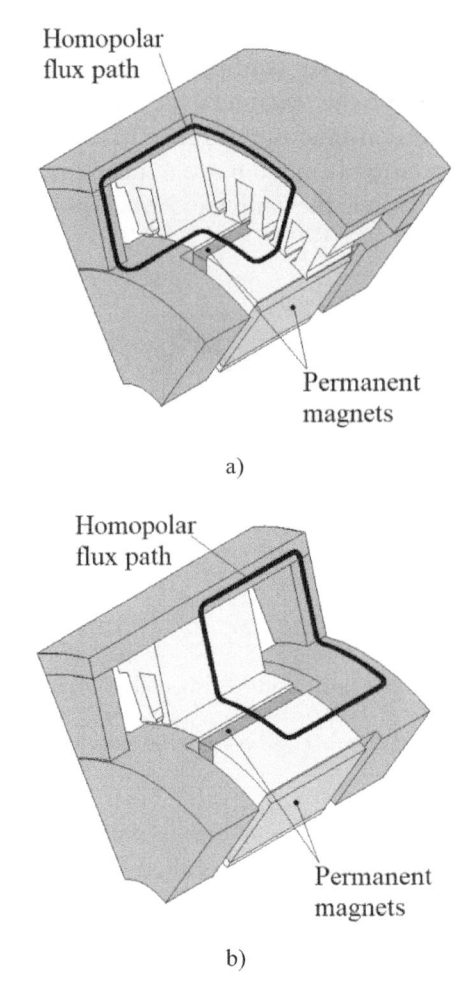

Figure 3.4. *Homopolar flux paths coming from permanent magnets*

For the bipolar configuration, the peak-to-peak amplitude of the excitation magnetic flux that can be achieved when the excitation flux is increased is greater than that for the homopolar configuration.

For the homopolar configuration, as the excitation coils act only on one type of magnetic pole, the increase in excitation flux is limited by the magnetic saturation of the pole concerned, whereas the adjacent pole may not be saturated and thus constitute a reserve of increase, but we do not have access to this.

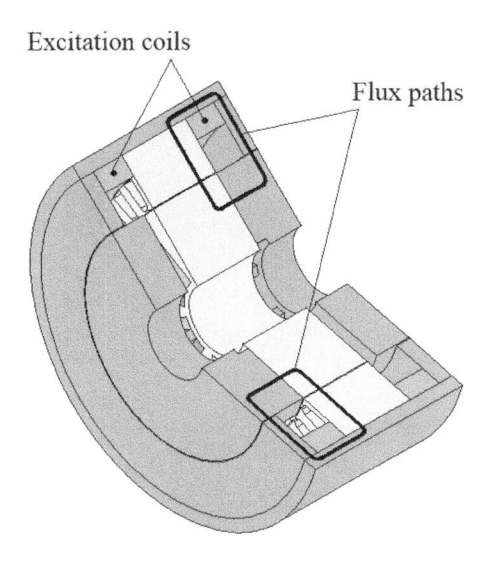

Figure 3.5. *Homopolar flux paths from the excitation coils*

For the bipolar configuration, since the excitation coils act on both types of magnetic poles (Figure 3.5), the excitation flux increase is limited by the magnetic pole saturation. Thus, it is possible to have a greater peak-to-peak amplitude of the excitation magnetic flux in the bipolar configuration than in the homopolar configuration. The experimental study of the no-load operation of this structure for both configurations will confirm this fact (section 3.2.3) (Vido et al. 2005a, 2005b).

3.2.2. Construction

In this section, we present some elements inherent to constructing the "Machine 1" prototype. Its structure is part of a series of structures built and studied to validate the hybrid excitation principle. These structures share the same stator (Figure 3.6).

Figure 3.6 presents some photos of the assembled stator with the field coils as well as the armature windings, and Figure 3.7 presents the main dimensions. The laminated part of the stator is made of standard laminations (iron-silicon (FeSi) laminations). Figure 3.8 presents a picture of the rotor sheet (Figure 3.8(a)), and the rotor assembled in its bipolar configuration (Figure 3.8(b)).

160 Hybrid Excited Synchronous Machines

a)

b)

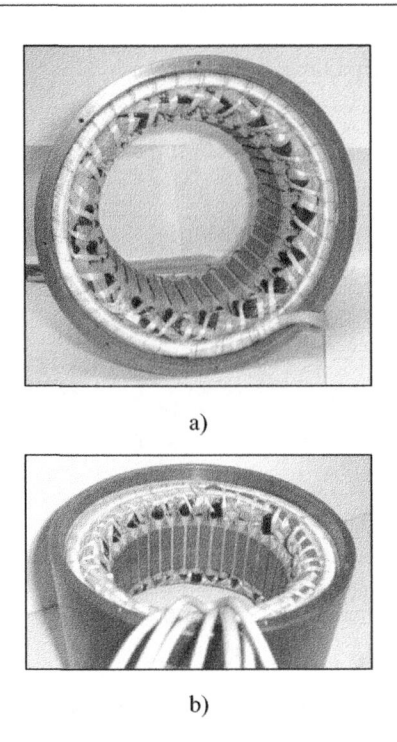

Figure 3.6. *Photos of the stator of "Machine 1"*

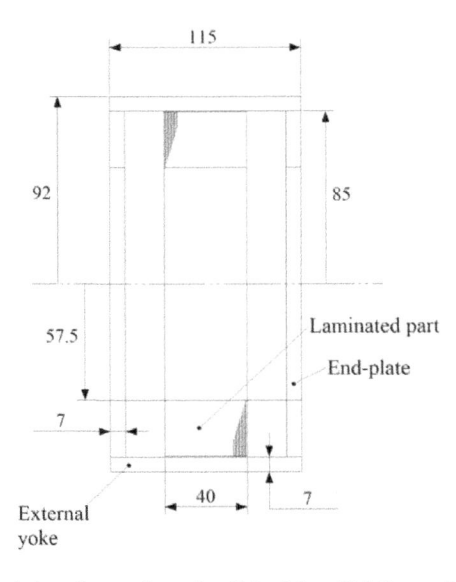

Figure 3.7. *The stator dimensions for "Machine 1" (dimensions in millimeters)*

Experimental Studies of Hybrid Excited Synchronous Machines 161

a) Rotor plate.

b) Rotor assembled in a bipolar configuration.

Figure 3.8. *Construction of the rotor*

Compared to the bipolar configuration (Figure 3.8(b)), the homopolar configuration is obtained by rotating one of the two rotoric flux collectors by one pole pitch.

Figure 3.9 presents a picture of the assembled machine, mounted on its test bench. It is connected to an alternator which serves as the load machine for the drive operation. We can see in this picture that the position sensor mounted at the end of the shaft is on the left of the machine, as well as the force sensor which is used to measure the torque (balance assembly). This bench was used to study the motor operation of this machine.

Figure 3.9. *Machine mounted on its test bench*

3.2.3. *Experimental study*

This structure was first tested under open-circuit conditions to evaluate its excitation flux control by using wound excitation. It was mounted on a lathe and driven at a speed of 170 rpm.

These tests concerned the structure in its homopolar and bipolar configurations (Machine 1). Figure 3.10 shows the variation curves of the excitation flux collected by the phases of the armature as a function of the excitation current for the two configurations. These measurements, as well as those made with other machines (Vido et al. 2005a, 2005b), confirm the possibility that the bipolar configuration would provide a wide range of excitation flux control, rather than the homopolar configuration.

Figure 3.11 shows the EMF (electromotive force) waveforms for "Machine 1" at different excitation current values I_{exc}, which itself was estimated by integrating these curves with respect to time.

Due to a wider excitation flux control range in the bipolar configuration (Machine 1), the motor operation tests were performed with the latter.

Figure 3.12 shows the schematic diagram of the test bench with its power supply. Figures 3.13 and 3.14 show the steps taken to study the operation on the whole "torque/speed" plane (Amara 2001; Amara et al. 2004a, 2009, 2021a).

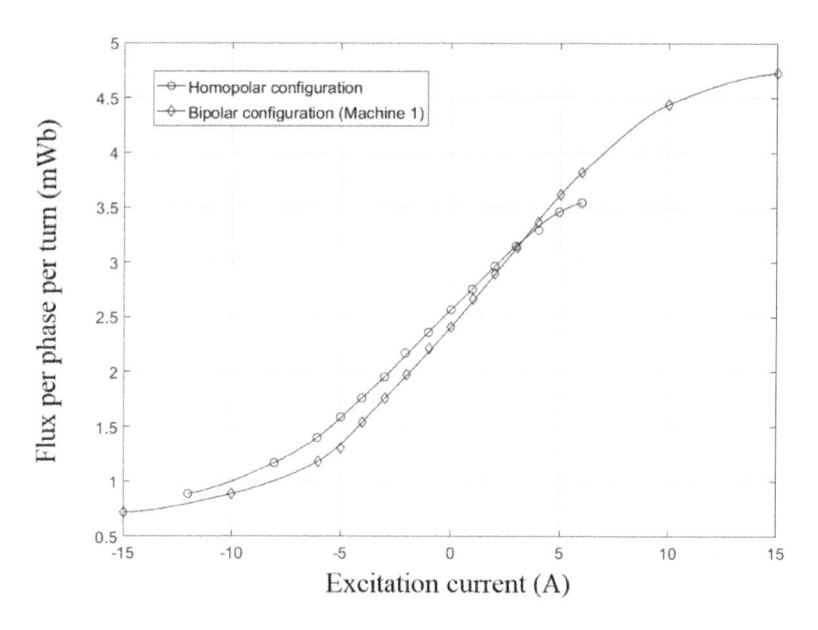

Figure 3.10. *Excitation flux variations for the two configurations*

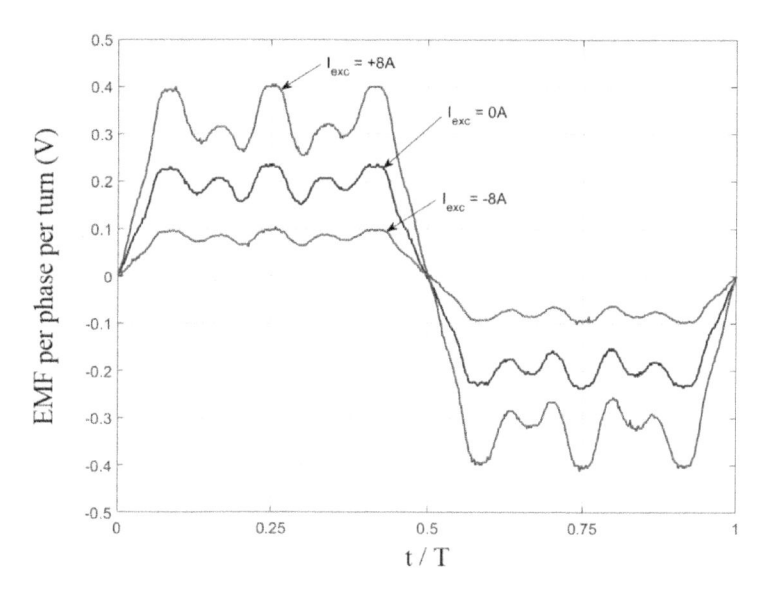

Figure 3.11. *EMFs for different excitation current values (N = 170 rpm)*

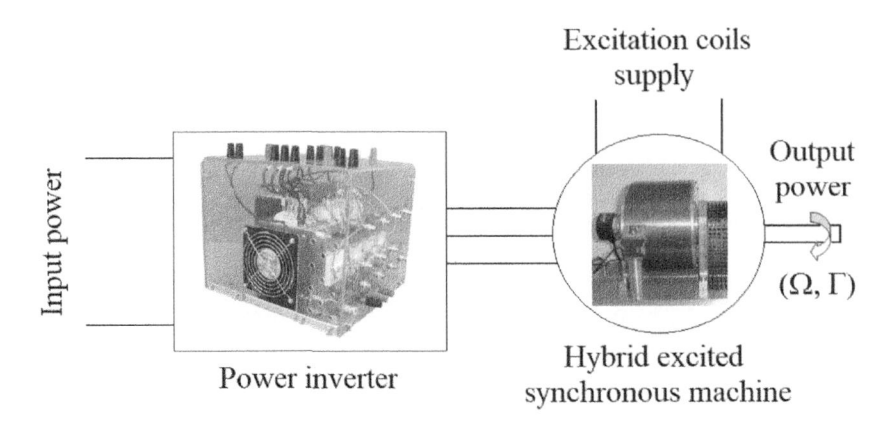

Figure 3.12. *Schematic diagram of the test bench in motor operation*

The efficiency is estimated by dividing the mechanical power output by the sum of the powers supplied to the hybrid excited synchronous power inverter assembly (Figure 3.12). The power supplied to the machine converter assembly includes the power supplied to the inverter and the field coils, respectively. The mechanical power is estimated from speed and torque measurements.

Figure 3.13 shows the test protocol followed when investigating the efficiency on the entire "torque/speed" plane. For speeds varying between 1,000 and 6,000 rpm in 1000 rpm steps, a reference armature current is set for the inverter. Four values of the armature current I_{aref} are set for each speed, i.e. 2.5, 5, 7 and 10 A. Five values of the excitation current I_{exc} are then imposed for each combination (speed, reference armature current), i.e. −4, −2, 0, 2 and 4 A. For each combination (reference armature and excitation currents), the "armature current/EMF" phase shift angle ψ that maximizes the output power (or torque) is identified. Once a triple ($I_{armature}$, I_{exc}, ψ) maximizing the output power (or torque) is identified, the efficiency is estimated for that point. A total of 120 measurement points (6 speed values × 4 reference armature current values × 5 field current values) that cover the entire torque/speed plane are generated.

Figure 3.14 illustrates the next step in the process during which the efficiency maps on the entire "torque/speed" plane are obtained. This second step starts by dividing the "torque/speed" plane into 72 regions (6 speed values × 12 torque values).

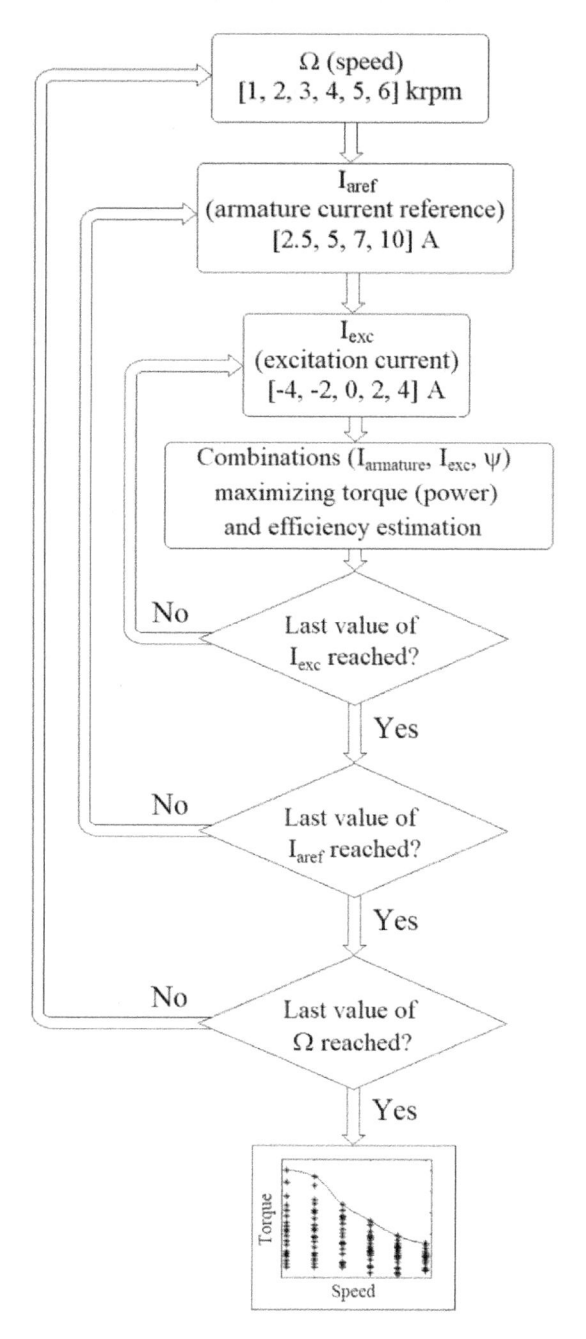

Experimental Studies of Hybrid Excited Synchronous Machines 165

```
                    Ω (speed)
                 [1, 2, 3, 4, 5, 6] krpm
                         ↓
                      I_aref
              (armature current reference)
                    [2.5, 5, 7, 10] A
                         ↓
                       I_exc
                 (excitation current)
                   [-4, -2, 0, 2, 4] A
                         ↓
            Combinations (I_armature, I_exc, ψ)
              maximizing torque (power)
                and efficiency estimation
                         ↓
        No      Last value of
         ←──── I_exc reached?
                         ↓ Yes
        No      Last value of
         ←──── I_aref reached?
                         ↓ Yes
        No      Last value of
         ←──── Ω reached?
                         ↓ Yes
                   [Torque vs Speed plot]
```

Figure 3.13. *Test protocol that estimates the performance over the entire "torque/speed" plane*

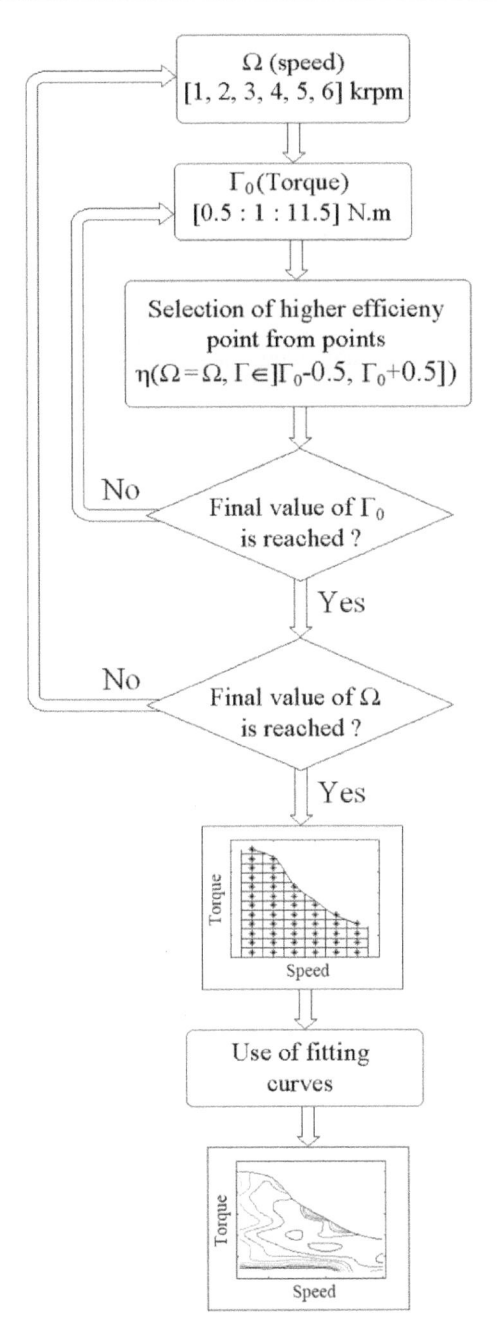

Figure 3.14. *Algorithm determining the maximum efficiency points in the "torque/speed" plane*

The speed reference is fixed by choosing the appropriate supply frequency. The angle ψ maximizing the output power (or torque) is identified by adjusting it to detect which value separates the regions in which the machine accelerates from the one in which it decelerates. The converter assembly machine is controlled by a dSPACE control board (Amara 2001) and, in addition to position and torque sensors, current sensors were used to control the assembly. The currents were controlled by means of a hysteresis control.

In each area of the "torque/speed" grid plane, the algorithm in Figure 3.14 allows us to select the operating point with the greatest efficiency. Finally, using interpolation curves allows for efficiency maps to be generated in the "torque/speed" plane.

The procedure shown in the two figures (Figures 3.13 and 3.14) is applied to "Machine 1" for two values of the hybridization ratio (section 2.4.1.2), $\alpha \approx 0.72$ and $\alpha = 1$. For $\alpha \approx 0.72$, the wound excitation is used to both increase the total excitation flux and to reduce it. In this case, we have 120 measurement points. For $\alpha = 1$, the wound excitation is used solely to weaken the excitation flux. The number of measurement points is reduced compared to the case $\alpha \approx 0.72$, for which there were 72 measurement points generated. Since the excitation current is used to reduce just the excitation flux of the magnets, the only values of the excitation current, which were less than or equal to 0 A, were fixed at −4, −2 and 0 A.

Figures 3.15 and 3.16 show the efficiency maps for both cases. To simulate the study conducted in Chapter 2 (section 2.5.1), reduced values were adopted. The basis used to define these reduced values was slightly different from that adopted in Chapter 2 (section 2.4.1.2). For both cases, the speed was reduced with respect to a reference speed of 2,000 rpm (not necessarily the base speed), and the torque was reduced with respect to the maximum value reached. Even if these were not the same standardization base, the use of these reduced values makes it possible to verify whether the trends observed following the theoretical study in Chapter 2 (section 2.5.1) were respected by this experimental study or not.

Figure 3.15 shows two mappings for the case $\alpha \approx 0.72$, one with the experimental efficiency values (Figure 3.15(a)) and another with the excitation current values (Figure 3.15(b)), plotted as asterisks. The torque is normalized to a value of 11.4 N·m.

Figure 3.16 shows the equivalent mappings for the $\alpha = 1$ case. The torque this time is normalized to a value of 8.6 N·m.

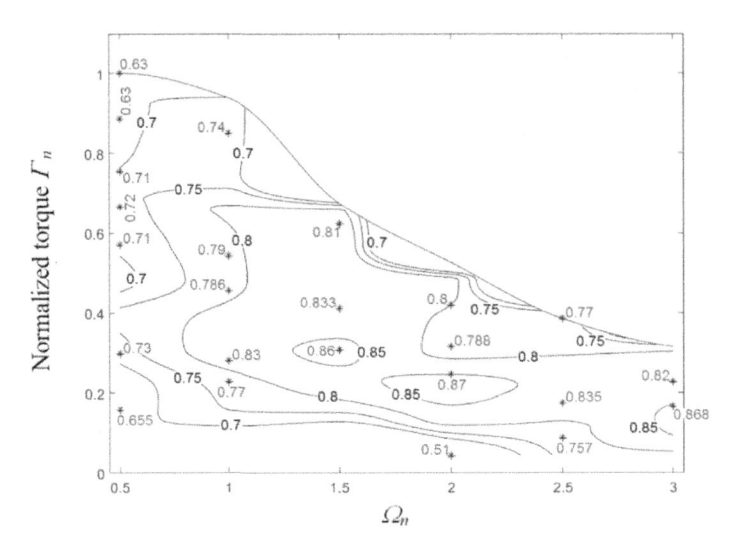

a) Efficiency map with the experimental efficiency values (asterisks).

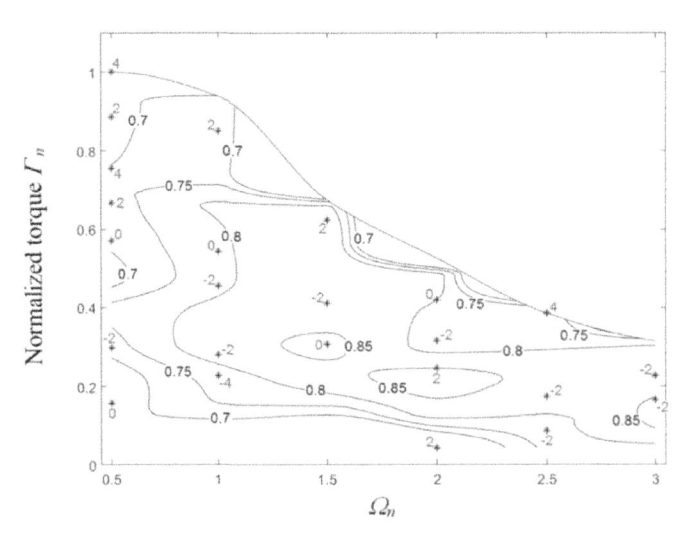

b) Efficiency map with excitation current values (asterisks).

Figure 3.15. *Efficiency maps in the case $\alpha \approx 0.72$*

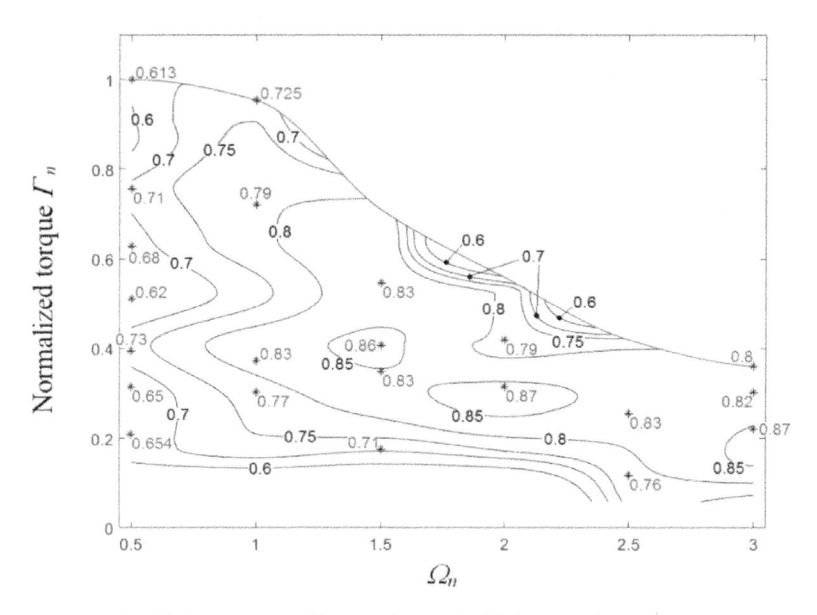

a) Efficiency map with experimental efficiency values (asterisks).

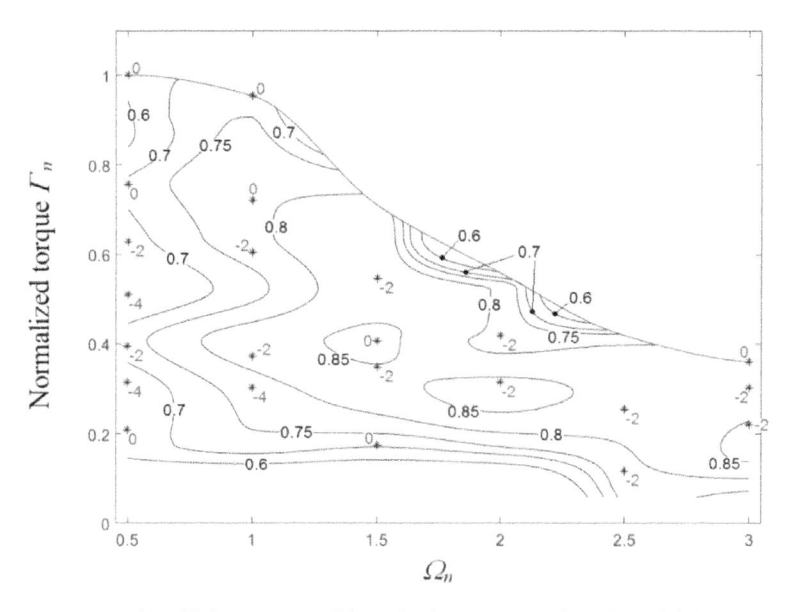

b) Efficiency map with excitation current values (asterisks).

Figure 3.16. *Efficiency maps in the case α = 1*

For a relatively low speed of $\Omega_n = 1.5$ (3,000 rpm), it should be noted that the highest efficiency point was obtained for $I_{exc} = 0$ A in both cases. On the other hand, Table 3.2 summarizes the distribution of excitation current values when the three highest efficiency values for each speed (18 steps) are considered. We note that the excitation current values for these points are largely distributed around the zero value.

I_{exc}(A)	-4	-2	0	2	4
$\alpha \approx 0.72$	3/18	6/18	4/18	4/18	1/18
$\alpha = 1$	3/18	7/18	8/18	0	0

Table 3.2. *Distribution of the excitation current values amongst the three highest efficiency values at each speed (18 steps)*

Comparing Figures 3.15 and 3.16, we also note that the areas with the greatest efficiencies are shifted towards the higher normalized torque values when the hybridization ratio is equal to 1 ($\alpha = 1$), compared to $\alpha \approx 0.72$. For a speed of 3,000 rpm ($\Omega_n = 1.5$), the highest efficiency point was shifted from the normalized torque value of ≈ 0.3 to ≈ 0.4, when the hybridization ratio was equal to ≈ 0.72 and 1, respectively. These results validate the theoretical study presented in Chapter 2 (section 2.5).

3.3. Machine 2

This structure was developed at the GREAH laboratory (Diab et al. 2021; Hlioui et al. 2022), designed and built within the framework of a project financed by the Normandy region (http://neptune-project.com/) and designed for a renewable energy application. For example, it could be used as a generator in wind or water turbines (http://bergey.com/wp-content/uploads/excel-10-owners-manual-2.pdf). It is a hybrid excited flux-switching synchronous machine with a Vernier effect (Chen et al. 2008; Zhu et al. 2008; Xie and Xu 2010; Diab et al. 2021). Vernier effect machines are often well suited for low-speed, high-torque machine applications (Lee 1963; Mailfert and Sargos 2004), which are characteristic of renewable energy applications. However, Vernier machines often suffer from low power factors (Diab et al. 2021).

Figure 3.17 shows a 3D view of the structure of "Machine 2". The coil heads are not shown. All the excitation sources, permanent magnets, excitation coils, as well as the armature windings are located in the stator. The rotor is completely passive and has 38 teeth. Since this machine has an even number of rotor teeth, the unbalanced magnetic force is theoretically null, provided that the rotor is well centered.

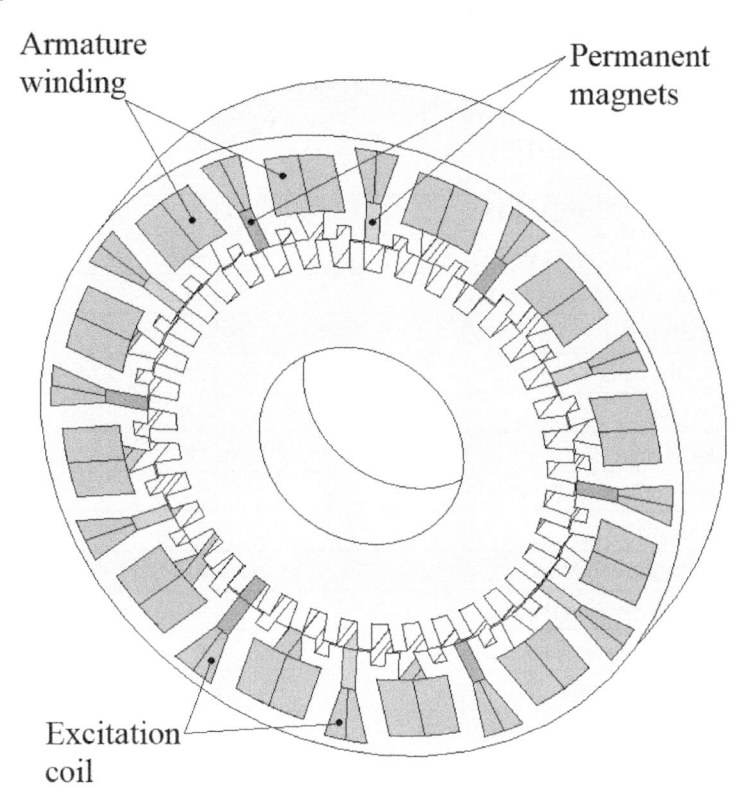

Figure 3.17. *3D structural view of "Machine 2"*

Hybrid excitation is mainly used to address the issue of availability and cost of rare-earth elements, as well as to increase the fault tolerance of "Machine 2".

Table 3.3 lists some important features of the structure, and additional information is provided in the following sections.

The structure of this machine was inspired by work done on hybrid excited flux-switching machines (Hoang et al. 2007; Gaussens 2013), as well as those including the Vernier effect (Chen et al. 2008; Zhu et al. 2008). A similar structure has been the subject of a Chinese patent (see Xie and Xu 2010).

Such structures can be considered to be original, and there is a distinct lack of studies on them. In addition, the structure studied in this section also includes damping windings used for induced voltage attenuation in the excitation coils (Wu et al. 2019a, 2019b, 2020; Hlioui et al. 2022), making it an original and unique structure. This solution was developed in the SATIE laboratory (Hlioui et al. 2022).

Parameter	Value
External diameter of the stator (mm)	400
Internal diameter of the stator (mm)	242
Air-gap thickness (mm)	1
Axial length of the machine (mm)	235
Active length (mm)	120
Number of pole pairs	38
Number of slots "armature"/"excitation"	12/12
Number of phases	3
Number of excitation coil turns	31
Number of armature winding turns	9
Number of damping winding turns	4
Excitation coil resistance (Ω)	1.1
Armature winding resistance per phase (Ω)	0.1
Damper winding resistance (Ω)	0.25
Permanent magnet characteristics (B_r, μ_r)	NdFeB (1.25, 1)
Magnet dimensions: thickness, height, length (mm)	8, 28, 120

Table 3.3. *Main characteristics of "Machine 2"*

3.3.1. *Structure and operating principle*

This machine can be described as a hybrid excited flux-switching machine with a Vernier effect, and also as a parallel hybrid excited machine with essentially a 2D structure. The ferromagnetic parts of the rotor and stator can be fully laminated.

Figure 3.18 shows the main excitation flux paths from the permanent magnets. It illustrates the operating principle of flux-switching (Hoang et al. 1997) in the 2D part of the machine. Depending on the relative position of the rotor with respect to the stator, the armature coil sees a flux that can be positive (Figure 3.18(a)) or negative (Figure 3.18(b)). It is therefore possible to induce flux variations in the armature coils, and consequently an EMF. This flux participates in torque generation (the principle of flux-switching for multiphase machines with a 2D structure was first discussed in work by Hoang et al. (1997)).

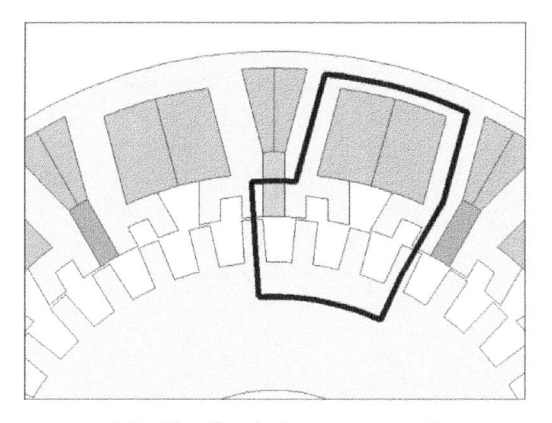

a) Positive flux in the armature coil.

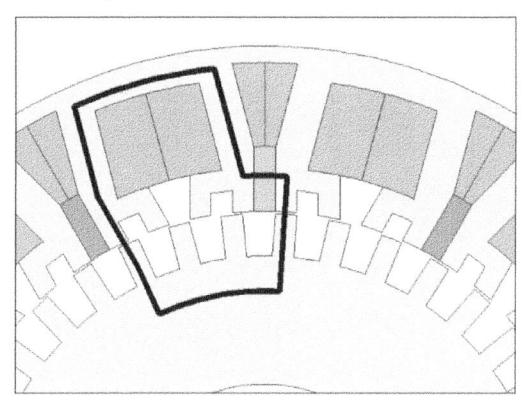

b) Negative flux in the armature coil.

Figure 3.18. *Operational principle of flux-switching (Hoang et al. 1997)*

Figure 3.19 shows another path that also comes from the permanent magnets. It can be described as leakage flux since it does not participate in any torque production. This path does not pass through the armature windings.

Figure 3.20 shows the main excitation flux paths from the permanent magnets and the excitation coil, with the same relative position of the rotor to the stator as in Figure 3.18(a). This figure illustrates how the wound excitation flux either consolidates the excitation flux of the permanent magnets or weakens it. The direction of the excitation current must be reversed depending on whether we wish to strengthen or weaken the excitation flux of the permanent magnets. Figure 3.21 is a copy of Figure 3.20, though now including the leakage flux from the permanent magnets as in Figure 3.19.

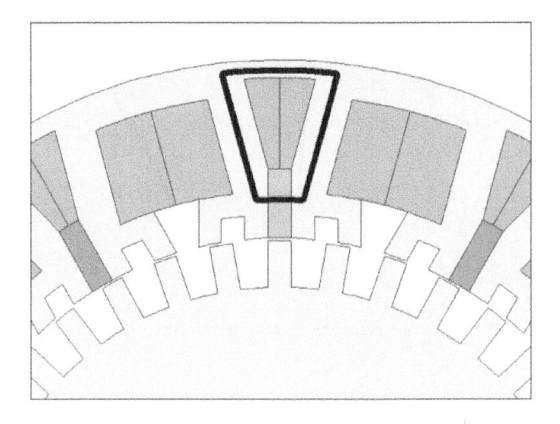

Figure 3.19. *Leakage flux coming from the permanent magnets*

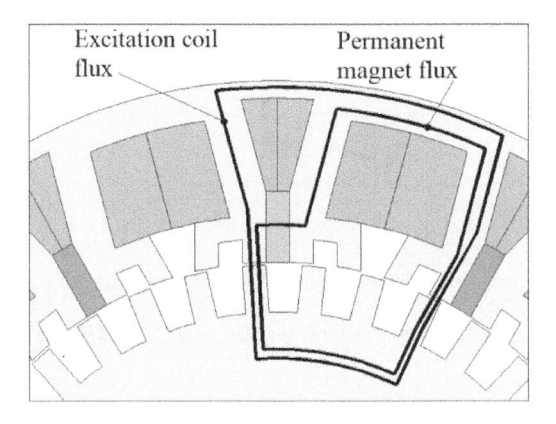

Figure 3.20. *Active flows coming from the two excitation flux sources*

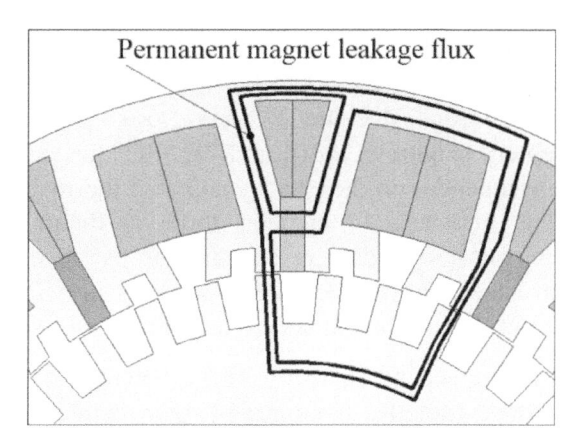

Figure 3.21. *The effect of the magnetic leakage flux on the total excitation flux variation*

Figure 3.21 allows us to anticipate the variation curve for the maximal total excitation flux as a function of the excitation current, which saturates more quickly when the excitation flux from the permanent magnets weakens. Indeed, in this case, the leakage flux of the permanent magnets would strengthen the wound excitation flux, and thus induce the external yoke to become saturated. It would be possible to increase the thickness of the external yoke and therefore postpone the onset of magnetic saturation, but this would be at a significantly greater cost in enlarging the external diameter of the machine and its mass. Moreover, there would be an increase in the leakage flux of the permanent magnets.

Similar structures have been presented in a Chinese patent (Xie and Xu 2010), but important differences exist between it and in "Machine 2". In the work by Xie and Xu (2010), two structures are described:

1) a three-phase machine with six stator modules, and a rotor with 19 teeth (Figure 3.22(a));

2) a four-phase machine with eight stator modules, and a rotor with 26 teeth (Figure 3.22(b)).

"Machine 2" has 12 stator modules and a rotor with 38 teeth (Figure 3.17). Even though the machines in Xie and Xu's patent and "Machine 2" are both hybrid excited machines with Vernier effects, the following differences can be listed:

1) Compared to three-phase machines (Figure 3.22(a)), the ratio "number of stator modules – number of rotor teeth" differs for "Machine 2". These numbers are doubled for "Machine 2", with "6–19" for the three-phase machine in the patent (Figure 3.22(a)) and "12–38" for "Machine 2". This has important consequences on the electromagnetic, thermal and mechanical dimensioning. The unbalanced magnetic force is theoretically zero for "Machine 2", whereas it should be present for the patented machine. For a given mechanical speed, the fundamental frequency is twice for "Machine 2", which affects the size of the magnetic circuit and the losses.

2) Compared to the second structure in the patent (Figure 3.22(b)), the ratio "number of stator modules – number of rotor teeth" and the number of phases are both different for "Machine 2". While the unbalanced magnetic forces should be theoretically zero for these two structures, the consequences described above regarding the difference between the number of stator modules and rotor teeth apply. Furthermore, the difference in the number of phases between the two structures implies a structural difference in the power-electronics converters that would be associated with them.

Figure 3.23 shows how the three-phase windings and the wound field excitation coils are distributed within "Machine 2". The letters "A", "B" and "C" represent the three phases, and the letters "E_e" and "E_s" represent the inputs and outputs of the excitation field coils, respectively, all of which are connected in series. Two adjacent field coils are wound in opposite ways. If one field coil is wound in the forward direction, then the other two adjacent coils are wound in the reverse direction. The distribution of the three phases in this figure is made assuming that the rotor is rotating in the forward direction.

Table 3.4 lists the main specifications on which the design of "Machine 2" is based. These are specifications used at the GREAH laboratory as a reference when comparing several prototypes (Tiegna 2013; Diab et al. 2021; https://greah.univ-lehavre.fr/spip.php?article215). The origin of these specifications was a wind turbine commercialized by the Bergey company (Xie and Xu 2010), which is a small wind turbine designed for use in domestic applications (individual houses or small farms), and which can be connected to the electrical grid (Xie and Xu 2010).

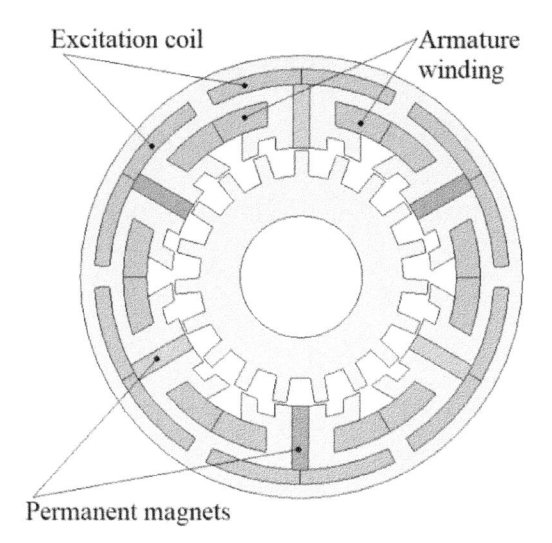

a) Structures 6–19 (three-phase armature).

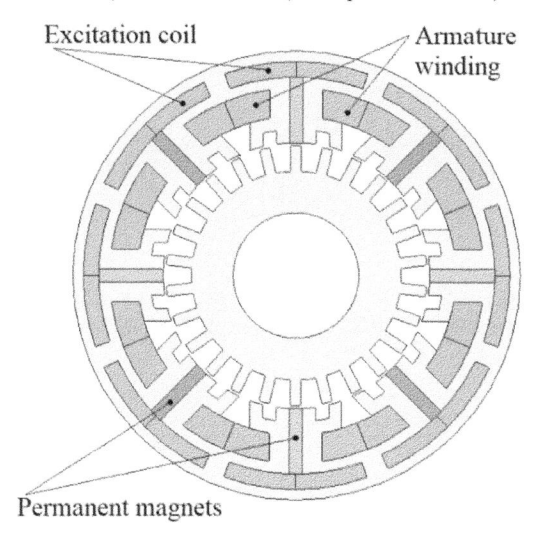

b) Structures 8–26 (four-phase armature).

Figure 3.22. *The structure of the patented machine (Xie and Xu 2010)*

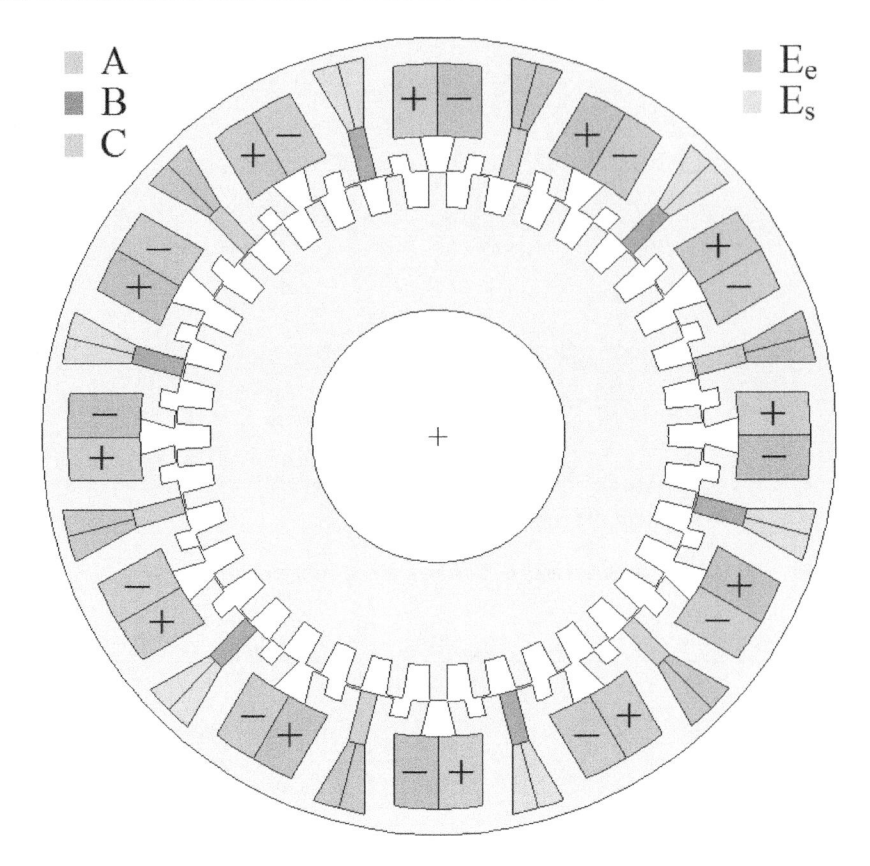

Figure 3.23. *The armature and excitation field winding distribution in the stator slots*

Parameter	Value
Nominal power (kW)	10
Nominal speed (rpm)	300
Rotor outer radius (mm)	200
Active length (mm)	120

Table 3.4. *Main specifications*

The dimensioning of "Machine 2" was carried out via parametric studies, with models based on finite element methods. As for "Machine 1", standard considerations from magnetic circuits, as well as inspiration from similar structures, allowed for a prototypical dimensioning to be established, before being refined with the aid of parametric methods. A detailed account of the performance study of the final machine (Machine 2) can be found in the work by Diab et al. (2021). The experimental study presented in Diab et al. (2021) is extended and is the topic of section 3.5.3.

It is relatively difficult to formalize the design and sizing of electrical machines, as it is both an art and a science (Lipo 2017). It is a multifactorial and iterative process, highly dependent on the experience of the designer. We therefore choose not to present all the steps leading to the final dimensions of "Machine 2".

3.3.2. Construction

Figure 3.24 shows the technical drawings and dimensions of the stator and rotor laminations (iron-silicon (FeSi) laminations). The internal radius of the rotor is 60 mm. Figure 3.25 presents the produced stator laminations, which were cut out by an EDM machine (wire cutting). Figure 3.26 presents the jig used to produce the armature windings, and Figure 3.27 presents the various windings and excitation coils assembled in the stator slots. The three phases of the armature are arranged in the star winding configuration.

In order to ensure better performance and stability, a water-cooling system was specifically designed for the prototype. It consists of a water jacket surrounding the stator since the main losses come from within it. Figure 3.28 presents some technical drawings of this jacket, where the water inlet and outlet can be seen (Figure 3.28(c)). Figure 3.29 presents CAD-generated views of the test bench, including the prototype as well as the drive machine (Leroy-Somer 2021). An estimate for the mass of the prototype indicated a value of 285 kg for the active parts (Diab et al. 2021). The base was designed to widen the base of the experimental bench, spreading out the mass of the machine over a greater area, to take into account the maximum load limit (< 500 kg/m^2) that the GREAH laboratory floor can support.

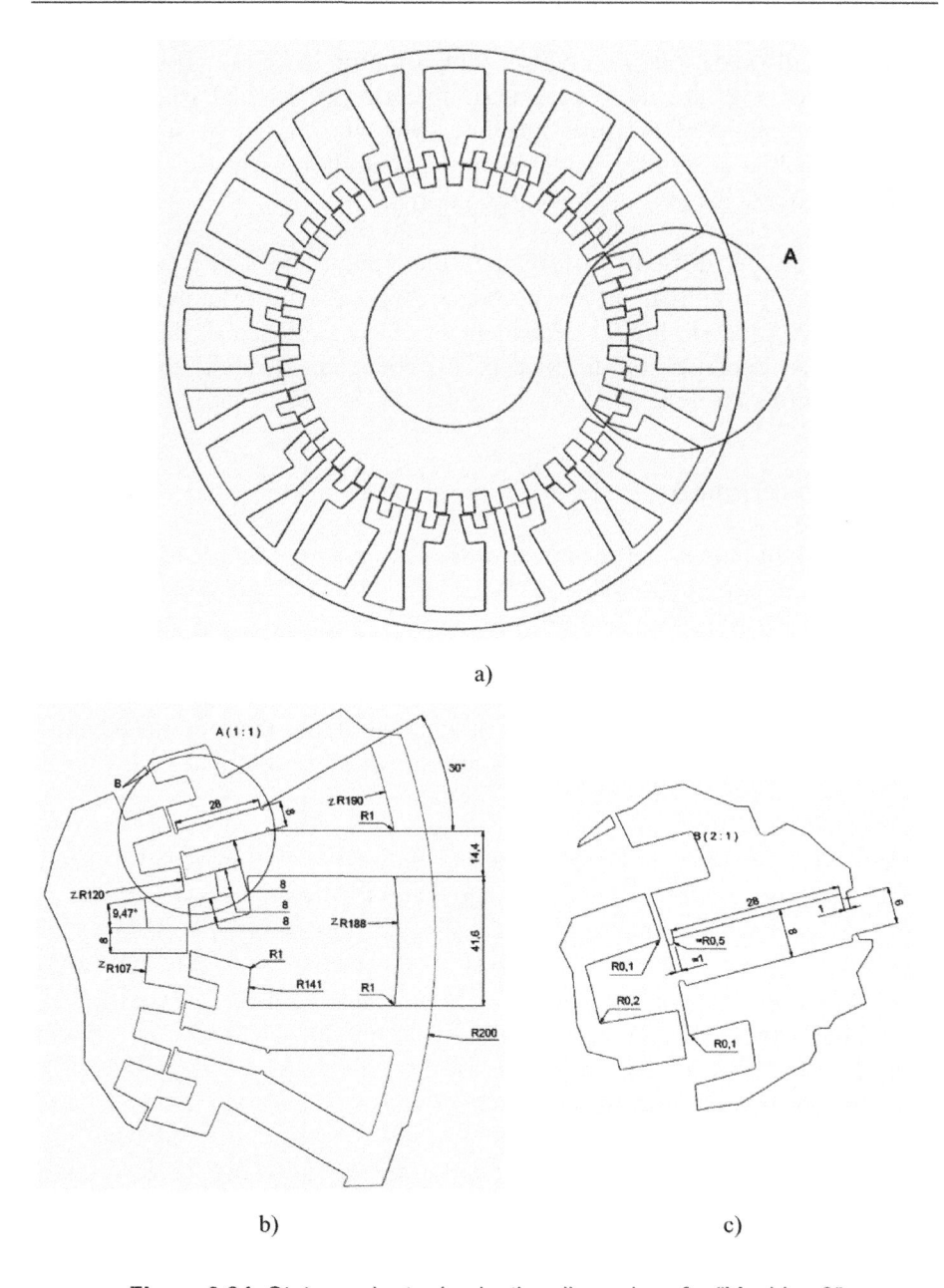

Figure 3.24. *Stator and rotor lamination dimensions for "Machine 2" (courtesy of the AE-Group, the Netherlands (https://ae-grp.nl/))*

a)

b)

c)

Figure 3.25. *Construction of the laminated stator stack (courtesy of the AE-Group, the Netherlands (https://ae-grp.nl/))*

182 Hybrid Excited Synchronous Machines

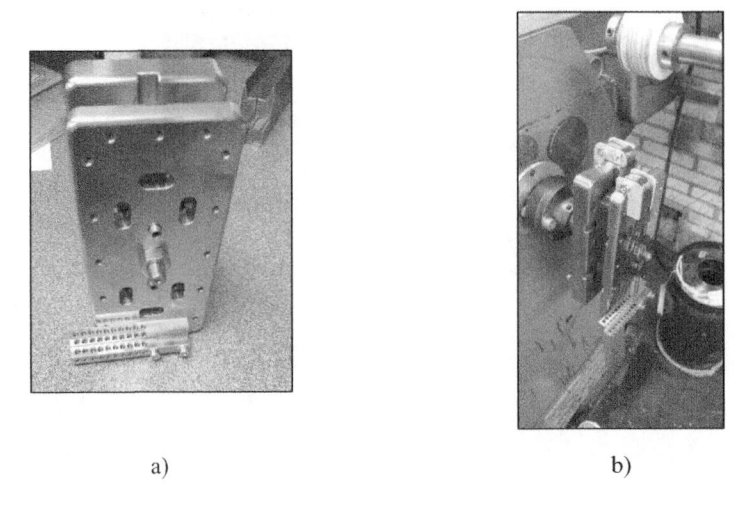

a)   b)

Figure 3.26. *Armature windings (reproduced with the kind permission of the AE-Group, the Netherlands (https://ae-grp.nl/))*

a) Front panel   b) Rear side

Figure 3.27. *Installation of the armature windings and excitation field coils in the stator slots (courtesy of the AE-Group, the Netherlands (https://ae-grp.nl/))*

Figure 3.30 shows the test bench base during its construction, which makes it possible to widen the base of the test bench, distributing it over 6 feet and reducing the weight per square meter. Indeed, the combined weight of the prototype and the gear motors used in the drive (Leroy-Somer 2021), along with that of the support table, exceeds the maximum admissible load, but the base allowed for the total load to be kept within the allowable range. Figure 3.31 presents a CAD-generated image, as well as a photograph of the terminal block of the machine.

a)

b)

c)

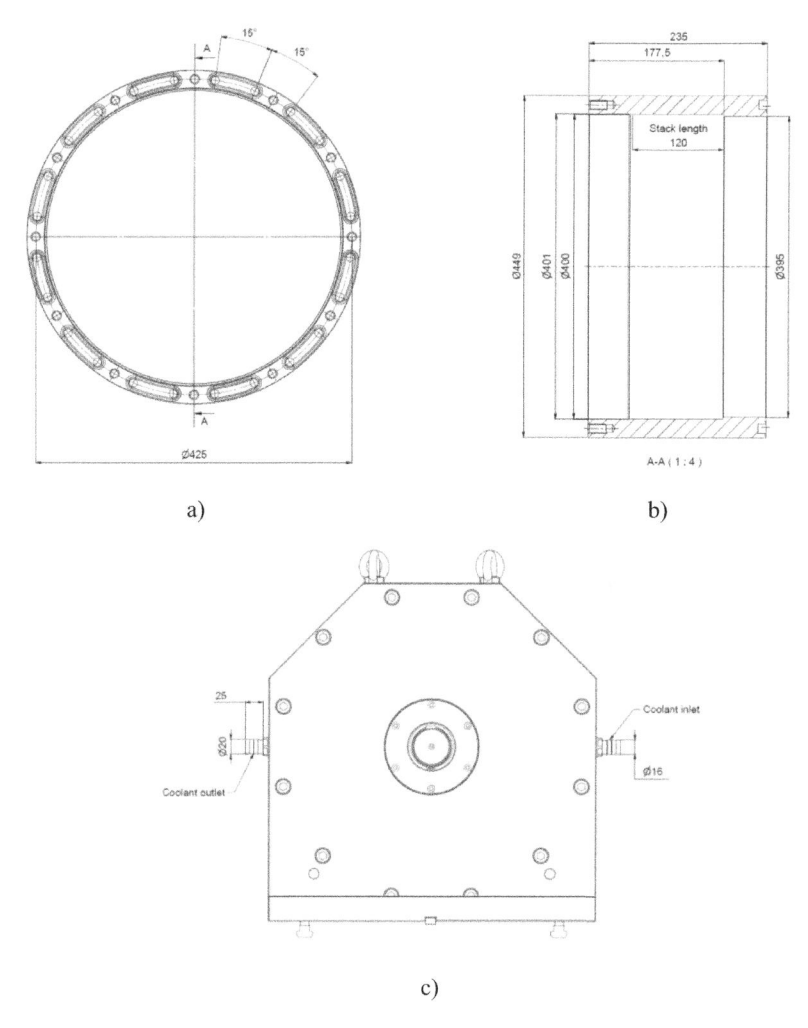

Figure 3.28. *Water-cooling system of the prototype (courtesy of the AE-Group, the Netherlands (https://ae-grp.nl/))*

a) 3D view of the test bench.

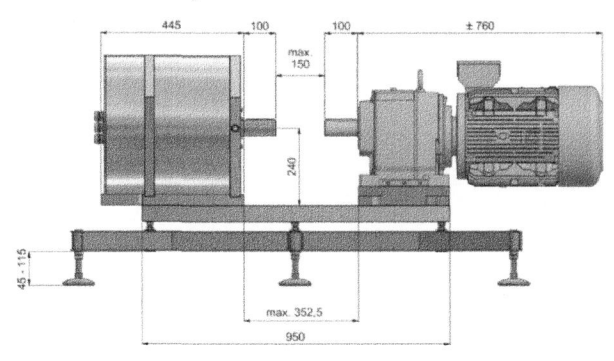

b) Side view of the test bench.

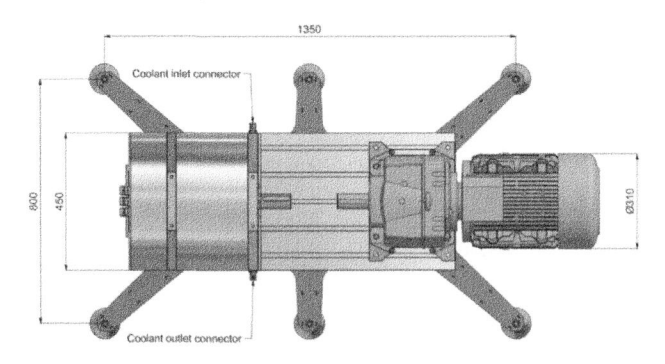

c) Top view of the test bench.

Figure 3.29. *CAD-generated views of the test bench (courtesy of the AE-Group, the Netherlands (https://ae-grp.nl/))*

Experimental Studies of Hybrid Excited Synchronous Machines 185

Figure 3.30. *Construction of the test bench base (courtesy of the AE-Group, the Netherlands (https://ae-grp.nl/))*

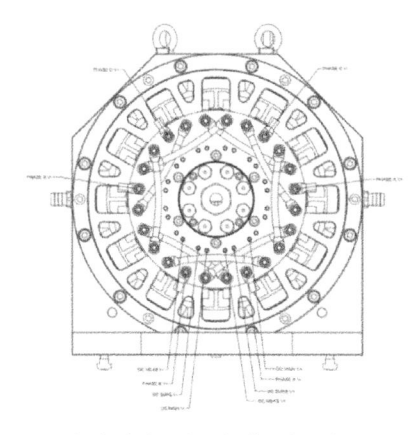

a) Technical drawing indicating the stator windings and coil connections.

b) Prototype back side with the stator windings and coil connections.

Figure 3.31. *Views of the back side of the prototype with the stator windings and coil connections (courtesy of the AE-Group, the Netherlands (https://ae-grp.nl/))*

Figure 3.32 shows the test bench installation in the GREAH laboratory. In this picture, we can see the different elements of the test bench, with a torque-sensor installed between the two machines (HBM), and the closed-loop water cooling system, just like for cars. The radiator that allows for the heat to be exchanged with the external environment is clearly visible. A future consideration is to combine this with a fan, to facilitate the exchange of heat. Lastly, it should be noted that the prototype is also equipped with a position and speed sensor (RM44D06_02.pdf).

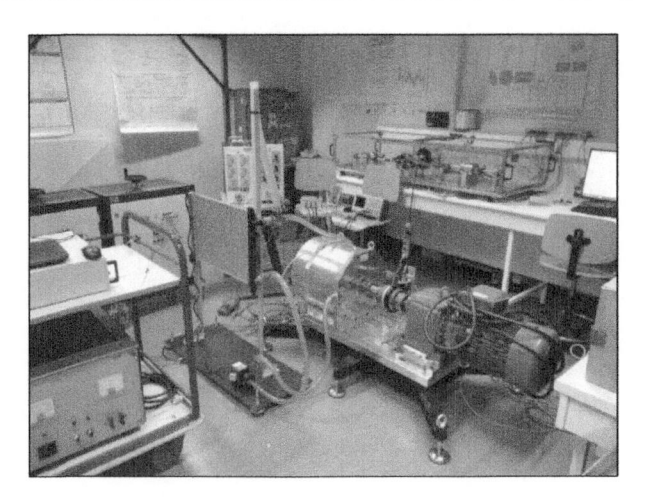

Figure 3.32. *"Machine 2" mounted on the test bench*

3.3.3. Experimental study

Like the previous structures, "Machine 2" has been tested at no-load and under-load conditions. In this section, we will reproduce these experimental tests which allows for the contribution that hybrid excitation can provide to renewable energy applications to be appreciated, which is unlike any previous applications. The experimental study will also validate the solution developed in the SATIE laboratory that reduces the induced voltage in the excitation coil (Hlioui et al. 2022).

For the no-load tests, the machine was driven at a speed of 342 rpm. Figure 3.33 compares the measured phase EMF waveforms for three values of the excitation current I_{exc} = −15 A, 0 A, and +15 A, and their harmonic contents. It can be seen that the wound coil excitation allows for the excitation flux to be controlled and that the waveforms are close to being perfectly sinusoidal. It should be noted that shorting or leaving the damping winding open does not influence the no-load EMF waveforms.

Figure 3.34 shows the variation of the phase EMF RMS value of one phase as a function of the excitation current. It can be seen that magnetic flux control is very effective. Compared to the null excitation current, the excitation flux increased by +300% when I_{exc} = 35 A, and was weakened by −50% when I_{exc} = −15 A.

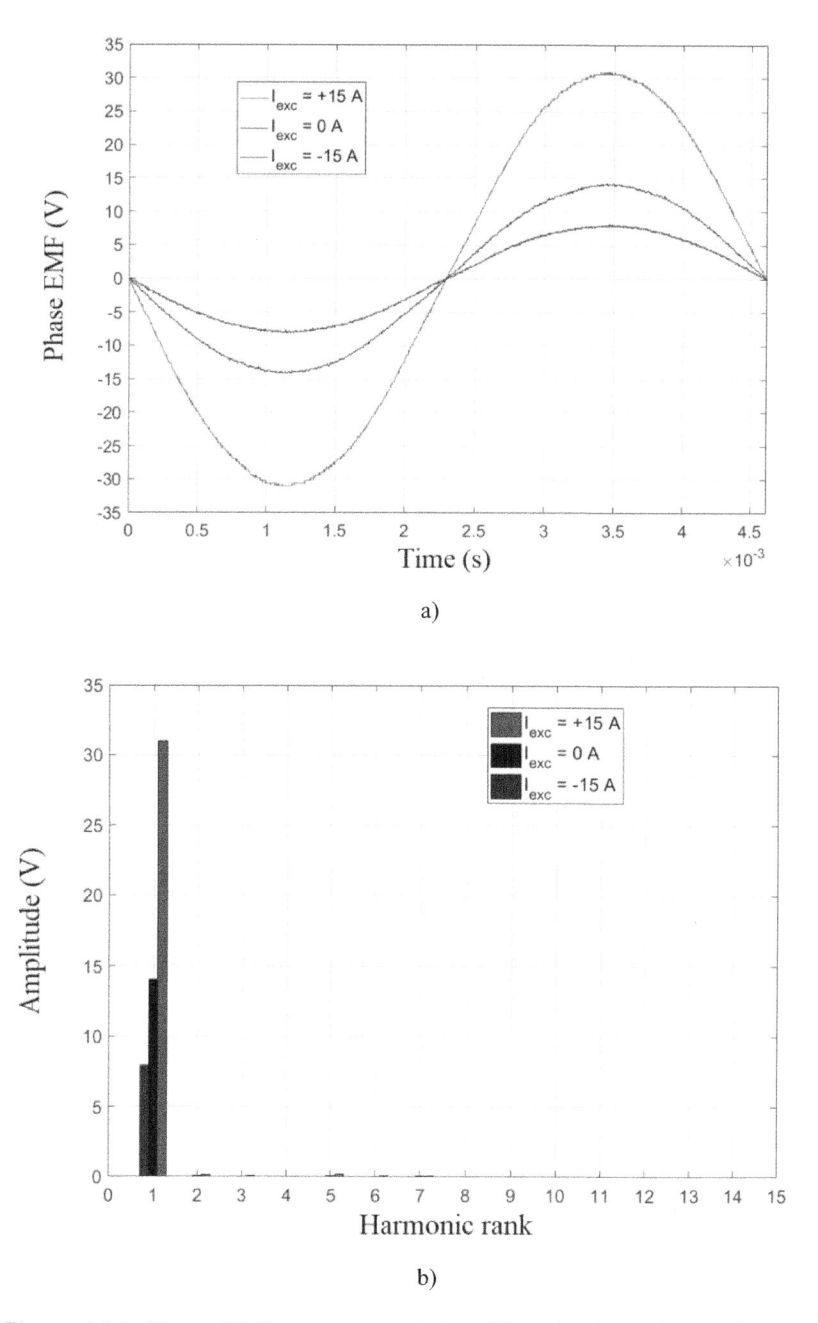

Figure 3.33. *Phase EMF measurements for different values of excitation current*

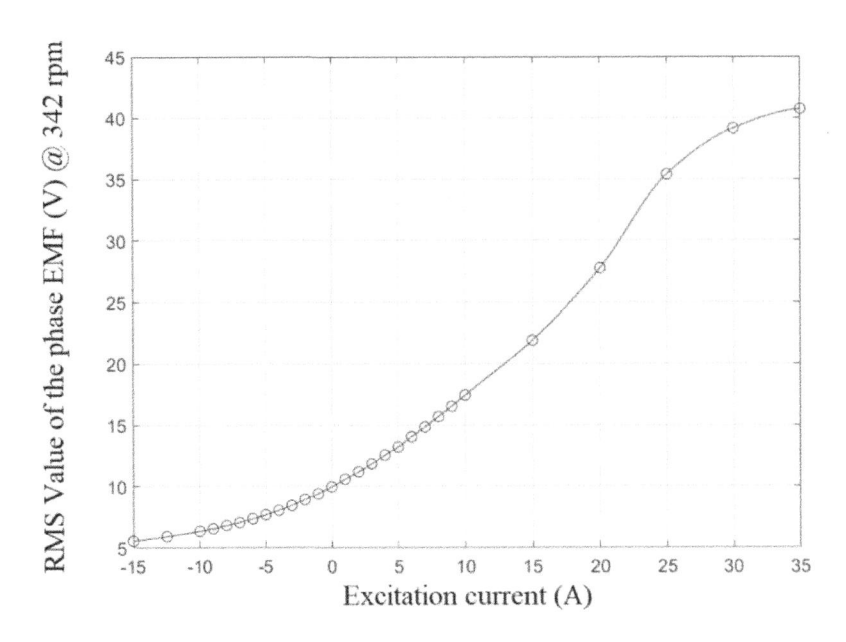

Figure 3.34. *Phase RMS value variation as a function of the excitation current*

Figure 3.35 shows the RMS value variation of the EMF induced in a measurement coil, which was wound in exactly the same way as the excitation coils and the damper windings. It can be seen that when short-circuited, the damper windings helped to reduce the induced EMF. Figure 3.36 shows the induced EMF variation in the measurement coil for $I_{exc} = 0$ A, when the damper winding was open and closed.

To study the load operation of "Machine 2", it was used as a generator (alternator) driven by an asynchronous machine via a gearbox (Figure 3.32) (Leroy-Somer 2021). The alternator delivered an inductive load consisting of four workloads (two resistive and two inductive), connected in parallel. Figure 3.37 shows the schematic diagram of the test bench. Taking into account the EMF level of the generator (Figure 3.34), and since there was no transformer to convert the voltage to that of the workloads, the tests were performed at partial load. During these tests, the alternator was driven at a speed of 342 rpm. Initially, the test consisted of an excitation current $I_{exc} =$ +20 A, a minimum resistance value for the resistive workloads, and a minimum inductance value for the inductive workloads. Figure 3.38(a) shows the simple phase voltage and current waveforms for one phase.

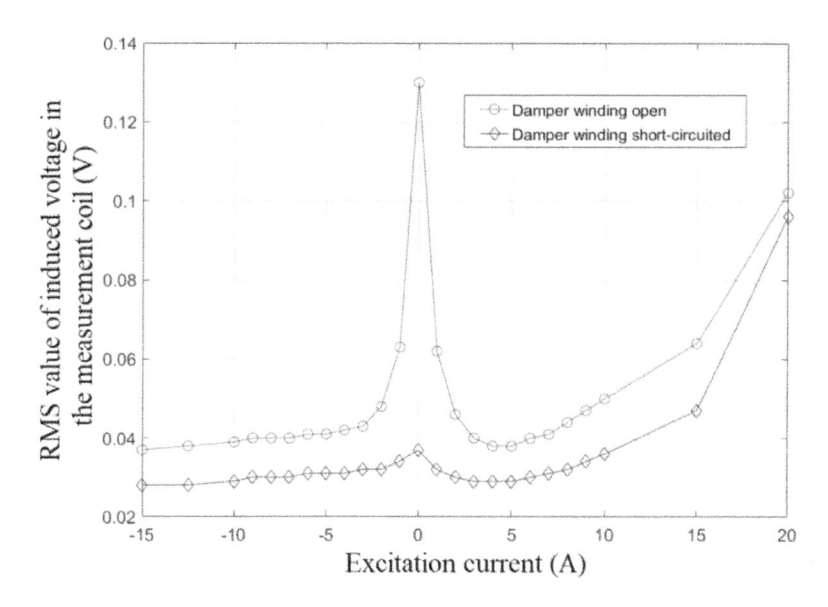

Figure 3.35. *RMS values of the induced voltage in the measurement coil*

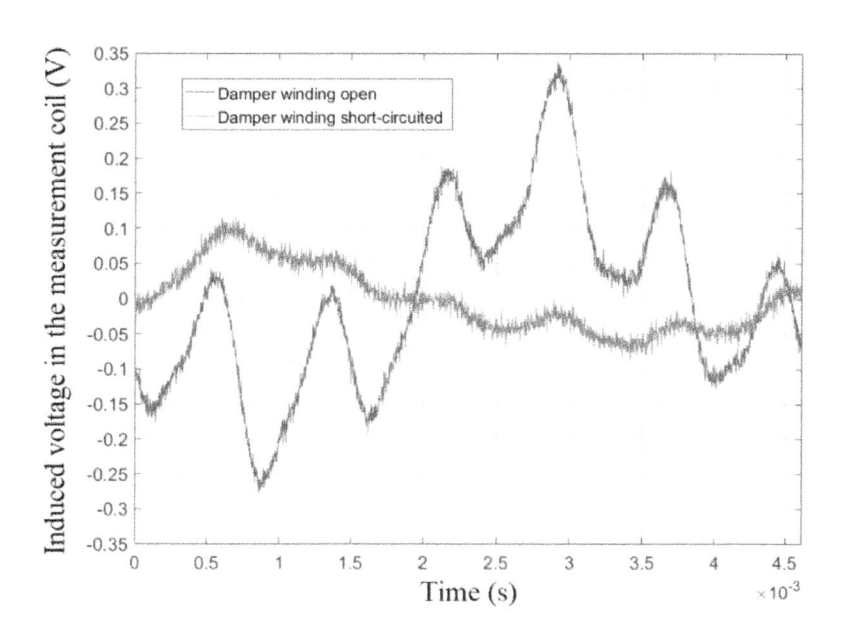

Figure 3.36. *Waveforms of the induced voltage in the measurement coil*

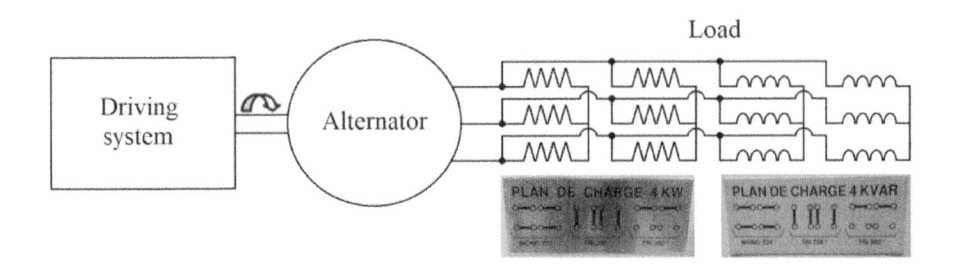

Figure 3.37. *Schematic diagram of the alternator test bench*

Afterwards, a maximum inductance value for the inductive workloads was imposed, and the excitation current was adjusted (I_{exc} = +18.8 A) to maintain the same line voltage value. The hybrid excitation then controlled the output voltage of the alternator. Figure 3.38(b) shows the waveforms of the line voltage and current for the same phase. When the inductance value is minimal (Figure 3.38(a)), the power factor is lower than that when the inductance value is maximal (Figure 3.38(b)). It should be noted that short-circuiting or leaving the damper windings open did not influence the waveforms of the alternator phase voltage.

Figure 3.39(a) and (b) compares the induced voltages in the measurement coil for the two tests. As with the no-load tests, it can be seen that short-circuiting the damper windings does reduce the induced voltage. Short-circuiting the damper winding essentially reduces the high-rank harmonics. The RMS voltage decreases overall from 113 mV (open damper winding) to 100 mV (short-circuited damper winding) for both tests. Other methods can be further adopted to reduce the induced voltage (Wu et al. 2019a, 2019b, 2020). Furthermore, it would be possible to combine these methods to remove the most significant harmonics from this voltage.

Figure 3.40(a) shows the RMS value variation in the short-circuited armature current as a function of the excitation current. Figure 3.40(b) shows the short-circuited current waveforms for three excitation current values I_{exc} = –15 A, 0 A and +15 A. The measurements are taken at a speed of 342 rpm. Note that short-circuiting or leaving the damper winding open does not influence the armature current waveforms.

a) Test with the minimal inductance value (I_{exc} = +20 A).

b) Test with the maximal inductance value (I_{exc} = +18.8 A).

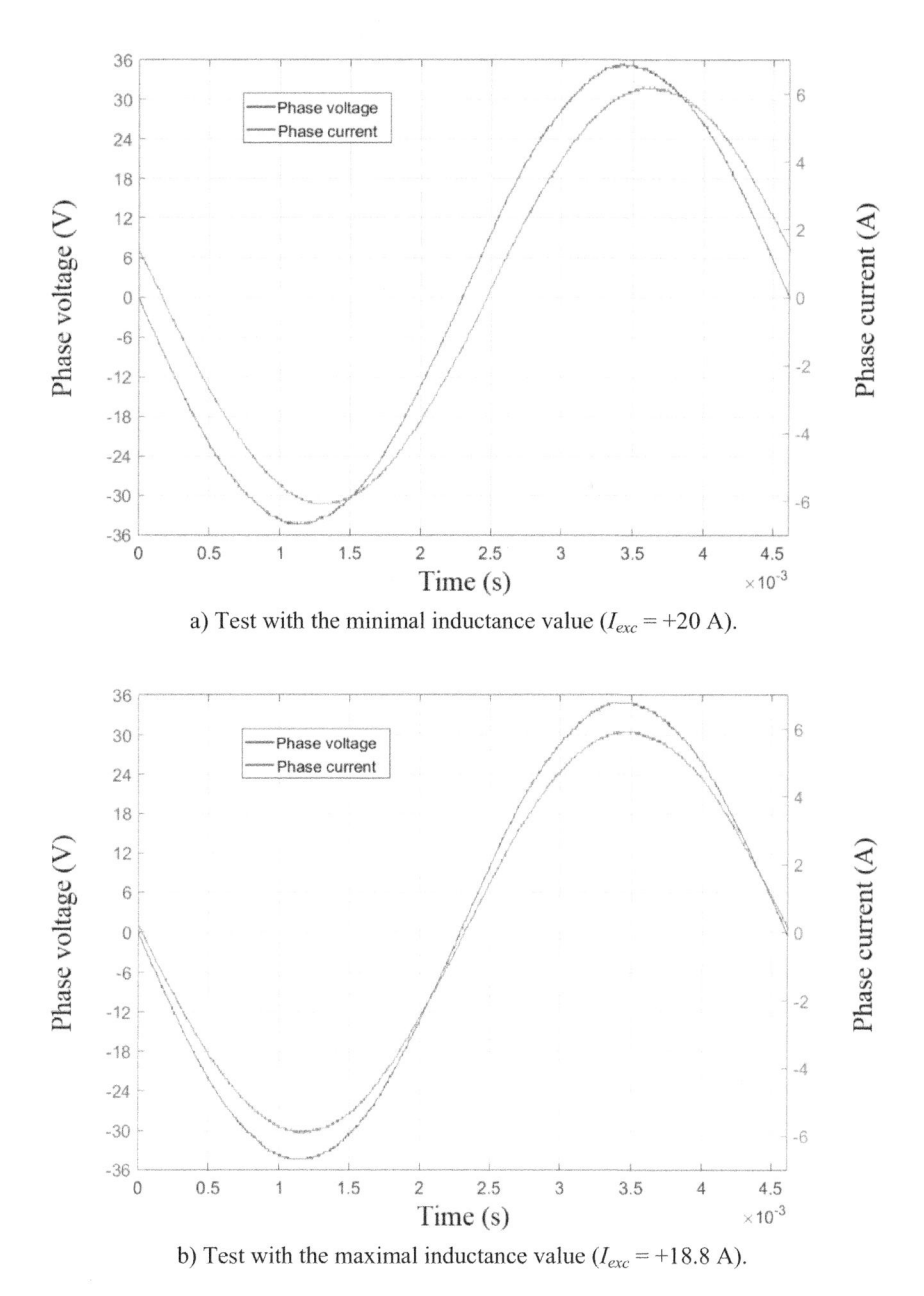

Figure 3.38. *Line voltage and current waveforms for test loads*

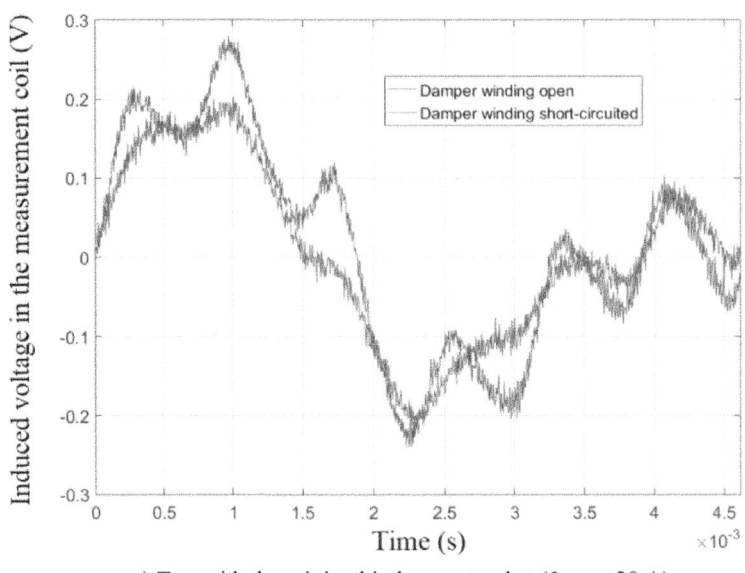

a) Test with the minimal inductance value ($I_{exc} = +20$ A).

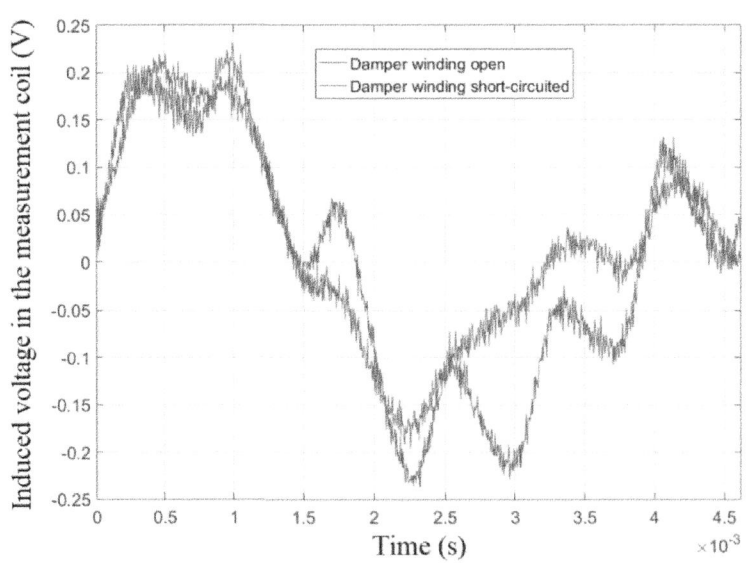

b) Test with the maximal inductance value ($I_{exc} = +18.8$ A).

Figure 3.39. *Induced voltage waveforms in the measurement coil for the load tests*

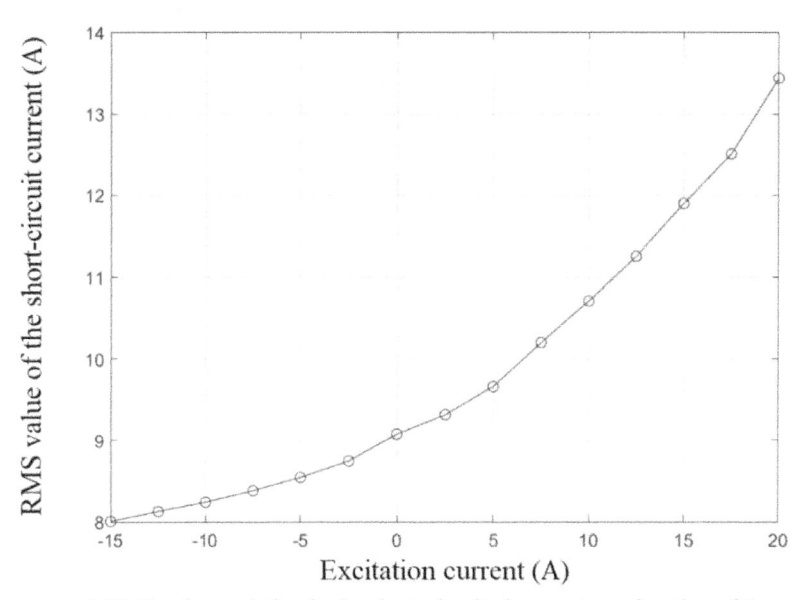

a) RMS value variation in the short-circuited current as a function of I_{exc}.

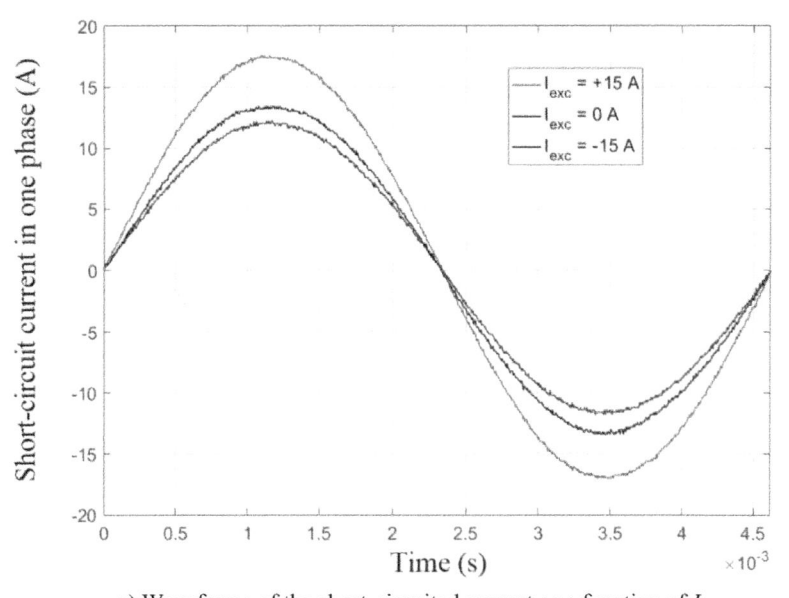

a) Waveforms of the short-circuited current as a function of I_{exc}.

Figure 3.40. *Short-circuited three-phase armature current at 342 rpm*

3.4. Conclusions and perspectives

This chapter, devoted to the experimental study of two hybrid excited machines, has allowed for the contribution that comes from hybrid excitation to be concretely illustrated. This study concerned itself with 3D machine ("Machine 1") and 2D machine ("Machine 2") structures for different applications. The motor and alternator operating modes were both addressed.

The study of "Machine 1" allowed us to present the hybrid excitation contribution towards extending the speed operating range in motor mode. In particular, the analysis of the motor operation on the entire "torque/speed" plane has allowed us to validate the study presented in section 2.5 of Chapter 2.

The generator operation (alternator) was studied using "Machine 2". It was designed to be used as a generator in renewable energy applications (wind or water turbines), at relatively low operating speeds (≈ 300 rpm). We were able to appreciate how the hybrid excitation principle allows us to control the alternator output values.

The machines studied in this chapter have a rotating structure. As a perspective for future research, it would be interesting to extend this study to linear machines. Figure 3.41 shows a 3D CAD view of a structure designed within the scope of the project, in which "Machine 2" was designed (http://neptune-project.com/).

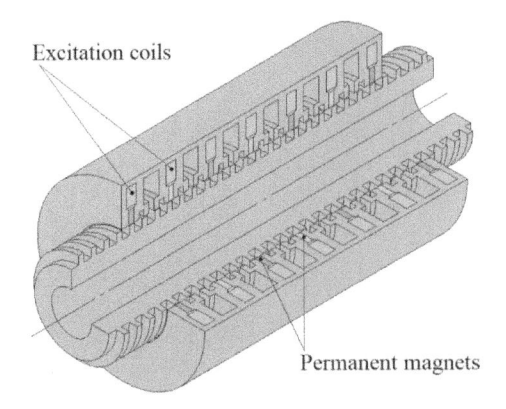

Figure 3.41. *Hybrid excitation flux-switching tubular linear machine with a Vernier effect*

It is a prototype of a direct drive wave-powered generator (Aubry et al. 2009; Ali 2021) and its structure can be considered to be the tubular linear version of "Machine 2". The armature windings are not shown in this figure.

Conclusion

This book has traced through the research work undertaken over more than two decades on hybrid excited synchronous machines (HESMs). We hope that it will resonate well within the community of electrical machine designers, and also more broadly with any researcher who is interested in the energy transition that is currently underway.

Several aspects inherent to designing and dimensioning HESMs have been addressed. The association of two excitation sources (permanent magnets and excitation coils) in synchronous machines allows for the design of some very varied structures.

Chapter 1 introduced the principle of hybrid excitation and discussed the state of the art of the resulting structures. Hybrid excitation offers additional degrees of freedom when it comes to designing and controlling HESMs, compared to other synchronous machines. The main degree of freedom was formalized by the so-called hybridization ratio, a parameter that allows for synchronous machines to be optimized, at the level of both the components and the system. HESMs are adaptable to many different applications and can be adjusted to satisfy various constraints in a flexible way. Different structures have been presented, and suitable classification criteria have been proposed.

Chapter 2 was dedicated to the operation of HESMs in the context of applications requiring the speed to vary. It has been shown, among other things, that adjusting the hybridization ratio allows for the energy efficiency to be optimized, in the context of an electric vehicle application. The more general subject of controlling synchronous machines was also discussed.

Indeed, in order to better situate HESMs, it was necessary to compare how they operate when compared to other synchronous machine technologies, including variable-reluctance synchronous machines. Reduced quantities were introduced, allowing for a general classification of synchronous machines in the plane (normalized inductance in the direct axis, saliency ratio) to be proposed. This chapter can be a good course material for electrical engineering students.

Finally, Chapter 3 was devoted to the experimental study of two machines with hybrid excitation. These studies allowed for the contribution that hybrid excitation provides in the operation of synchronous machines to be illustrated concretely. The prototypes studied were designed for different purposes; while the first prototype (Machine 1) is purely a laboratory-based structure designed to analyze the operation of HESMs, the second prototype (Machine 2) was designed for a more specific application. The studies have effectively validated the interest in HESMs and confirmed the flexibility that hybrid excitation provides via the additional degree of freedom.

In conclusion, we hope this book will be useful to both researchers and students of electrical engineering. We insist on the necessity to adopt eco-design approaches, not only considering the isolated component, but always to put things into perspective and take into account the nearby and far-off environments.

References

Advanced Electromagnets Group (2020). Home [Online]. Available at: https://ae-grp.nl/ [Accessed 9 November 2021].

Akemakou, A.D. (2006). Double-excitation rotating electrical machine for adjustable defluxing. U.S. Patent US20060119206A1.

Akemakou, A.D. and Phounsombat, S.K. (2000). Electrical machine with double excitation, especially a motor vehicle alternator. U.S. Patent US6147429.

Albert, L. (2004). Modélisation et optimisation des alternateurs à griffes. Application au domaine automobile. PhD Thesis, Institut National Polytechnique de Grenoble.

Amara, Y. (2001). Contribution à la conception et à la commande des machines synchrones à double excitation. Application au véhicule hybride. PhD Thesis, Université Paris XI.

Amara, Y. and Hlioui, S. (2019). Degrees of freedom in the design of PM synchronous machines. *Proc. 19th Int. Symp. Electromagn. Fields Mechatronics, Electr. Electron. Eng. (ISEF)*, Nancy, 1–6.

Amara, Y., Lucidarme, J., Gabsi, M., Lécrivain, M., Ben Ahmed, A.H., Akémakou, A.D. (2001). A new topology of hybrid excitation synchronous machine. *IEEE Trans. Ind. Appl.*, 37(5), 1273–1281.

Amara, Y., Ahmed, H.B., Gabsi, M., Lécrivain, M., Chabot, F. (2004a). Machines synchrones à double excitation : analyse et optimisation du fonctionnement pour la traction électrique. *RIGE*, 7, 163–199.

Amara, Y., Hoang, E., Gabsi, M., Lécrivain, M., Ben Ahmed, A.H., Dérou, S. (2004b). Measured performances of a new hybrid excitation synchronous machine. *EPE J.*, 12(4), 42–50.

Amara, Y., Gabsi, M., Vido, L., Ben Ahmed, A.H., Lécrivain, M. (2007). Machines synchrones à double excitation. Principes et structures. *RIGE*, 10, 151–188.

Amara, Y., Vido, L., Gabsi, M., Haong, E., Ben Ahmed, A.H., Lecrivain, M. (2009). Hybrid excitation synchronous machines: Energy-efficient solution for vehicles propulsion. *IEEE Trans. Veh. Technol.*, 58(5), 2137–2149.

Amara, Y., Gabsi, M., Vido, L., Ben Ahmed, A.H., Hoang, E., Lécrivain, M., Takorabet, A. (2010). Classification des alternateurs débitant sur un pont à diodes connecté à un bus continu. Caractéristique du débit maximal. *RIGE*, 13, 33–89.

Amara, Y., Hlioui, S., Belfkira, R., Barakat, G., Gabsi, M. (2011). Comparison of open circuit flux control capability of a series double excitation machine and a parallel double excitation machine. *IEEE Trans. Veh. Technol.*, 60(9), 4194–4207.

Amara, Y., Barakat, G., Paulides, J.J.H., Lomonova, E. (2013). Overload capability of linear flux switching permanent magnet machines. *Proc. 9th International Symposium on Linear Drives for Industry Applications (LDIA13)*, Hangzhou, 416–417, 345–352.

Amara, Y., Hlioui, S., Ben Ahmed, H., Gabsi, M. (2019). Power capability of hybrid excited synchronous motors in variable speed drives applications. *IEEE Trans. Magn.*, 55(8), 8204312.

Amara, Y., Hlioui, S., Ben Ahmed, H., Gabsi, M. (2021a). Pre-optimization of hybridization ratio in hybrid excitation synchronous machines using electrical circuits modelling. *Mathematics and Computers in Simulation*, 184, 118–136.

Amara, Y., Hlioui, S., Gabsi, M. (2021b). Overview of degrees of freedom in the design of PM synchronous machines. *Energies*, 14, 3990.

Ammar, A., Berbecea, A.C., Gillon, F., Brochet, P. (2012). Influence of the ratio of hybridization on the performances of synchronous generator with hybrid excitation. *Proceedings of 20th International Conference on Electrical Machines (ICEM 2012)*, Marseille.

Asfirane, S., Hlioui, S., Amara, Y., Gabsi, M. (2019). Study of a hybrid excitation synchronous machine: Modeling and experimental validation. *Math. Comput. Appl.*, 24(2), 34.

Aubry, J., Babarit, A., Ben Ahmed H., Multon, B. (2009). La récupération de l'énergie de la houle, partie 2 : Systèmes de récupération et aspects électriques. *La Revue 3E.I, Société de l'électricité, de l'électronique et des technologies de l'information et de la communication*, (69), 26–32.

Aydin, M. (2004). Axial flux surface mounted PM machines for smooth torque traction drive applications. PhD Thesis, University of Wisconsin-Madison.

Aydin, M., Huang, S., Lipo, T.A. (2002). A new axial flux surface mounted permanent magnet machine capable of field control. *Conf. Rec. IEEE IAS Annu. Meeting*, 2, 1250–1257.

Aydin, M., Lipo, T.A., Huang, S. (2007). Field controlled axial flux permanent magnet electrical machine. U.S. Patent US20070046124A1.

Barret, P. (1978). Machines synchrones. Fonctionnement en régime permanent. Dossier D480, Techniques de l'Ingénieur.

Ben Ahmed, H. (2006). Des procédés de conversion électro-magnéto-mécaniques non-conventionnels aux systèmes mécatroniques : Conception – Modélisation – Optimisation. HDR, Université Paris Sud – Paris XI.

Betz, R.E. (1992). Theoretical aspects of control of synchronous reluctance machines. *IEE Proceedings-B*, 139(4), 355–364.

Betz, R.E., Lagerquist, R., Jovanovic, M., Miller, T.J.E., Middleton, R.H. (1993). Control of synchronous reluctance machines. *IEEE Trans. Ind. Appl.*, 29(6), 1110–1122.

Biais, F. and Langry, P. (1998). Optimization of permanent magnet traction motor for electric vehicle. *Proc. of the 15th EVS*, Brussels, October.

Bianchi, N. and Bolognani, S. (1997). Parameters and volt-ampere ratings of a synchronous motor drive for flux-weakening applications. *IEEE Trans. Power Electron.*, 12(5), 895–903.

Bianchi, N., Bolognami, S., Chalmers, B.J. (2000). Salient-rotor PM synchronous motors for an extended flux-weakening operation range. *IEEE Trans. Ind. Appl.*, 36(4), 1118–1125.

Boldea, I. (2016). *Variable Speed Generators*, 2nd edition. CRC Press, Taylor & Francis Group, Boca Raton.

Boldea, I., Muntean, N., Deaconu, S., Nasar, S.A., Fu, Z. (1992). Distributed anisotropy rotor synchronous (DARSYN) drives-motor identification and performance. *Proceedings of the International Conference on Electrical Machines*, 542–546.

Bose, B.K. (2001). *Modern Power Electronics and AC Drives*. Prentice Hall, Hoboken.

Bürger, K.G. (1999). Alternateurs. *Cahier technique*, Robert Bosch GmbH.

Chalmers, B.J., Musaba, L., Gosden, D.F. (1996). Variable-frequency synchronous motor drives for electric vehicles. *IEEE Trans. Ind. Applicat.*, 32, 896–903.

Chao, Z., Yuefei, Z.U.O., Feng, L.I., Zixuan, X. (2018). Hybrid excitation direct-drive motor. Chinese Patent CN108336837A.

Chatelain, J. (1989). *Machines électriques Volume 10 : Traité d'électricité*. Presses Polytechniques Romandes, Lausanne.

Chen, J.T., Zhu, Z.Q., Howe, D. (2008). Stator and rotor pole combination and optimal design of multi-tooth flux-switching PM brushless AC machines. *IEEE Trans. Magn.*, 44(12), 4659–4667.

Christophe, B., Alain, S., Matthieu, J. (2014). Moteur synchrone avec contrôle de la magnétisation des aimants en fonctionnement. *Symposium de génie electrique, ENS de Cachan*, Cachan.

De Doncker, R., Pulle, D.W.J., Veltman, A. (2011). *Advanced Electrical Drives*. Springer, Berlin.

De La Barrière, O. (2010). Modèles analytiques électromagnétiques bi et tridimensionnels en vue de l'optimisation des actionneurs disques : étude théorique et expérimentale des pertes magnétiques dans les matériaux granulaires. PhD Thesis, École normale supérieure de Cachan.

Diab, H., Amara, Y., Hlioui, S., Paulides, J.J.H. (2021). Design and realization of a hybrid excited flux switching Vernier machine for renewable energy conversion. *Energies*, 14, 19:6060.

Diao, T. (2018). Alternating-current and permanent magnet hybrid excitation doubly-fed wind power generator and power generation system. Chinese Patent CN108282064A.

Dupas, A. (2016). Modélisation et optimisation d'une machine synchrone à commutation de flux et à double excitation à bobinage global. PhD Thesis, Université Paris-Saclay.

Dupas, A., Hlioui, S., Hoang, E., Gabsi, M., Lécrivain, M. (2016). Investigation of a new topology of hybrid-excited flux-switching machine with static global winding: Experiments and modeling. *IEEE Trans. Ind. Appl.*, 52(2), 1413–1421.

Fang, Y., Liu, Q., Ma, J., Meng, L., Wang, L., Xia, C., Yao, Y. (2010). Hybrid excitation structure. Chinese Patent CN101814821A.

Fernandez-Bernal, F., Garcia-Cerrada, A., Faure, R. (2001). Determination of parameters in interior permanent-magnet synchronous motors with iron losses without torque measurement. *IEEE Trans. Ind. Appl.*, 37(5), 1265–1272.

Fodorean, D., Djerdir, A., Viorel, I.-A., Miraoui, A. (2007). A double excited synchronous machine for direct drive application. Design and prototype tests. *IEEE Trans. Energy Convers.*, 22(3), 656–665.

Fukami, T., Hayamizu, T., Matsui, Y., Shima, K., Hanaoka, R., Takata, S. (2010). Steady-state analysis of a permanent-magnet-assisted salient pole synchronous generator. *IEEE Trans. Energy Convers.*, 25(2), 388–393.

Gaussens, B. (2013). Machines synchrones à commutation de flux : de la modélisation numérique et analytique à l'exploration topologique. PhD Thesis, École normale supérieure de Cachan.

Greif, H., Nguyen, N.-T., Mueller, A., Reutlinger, K. (2011). Electric machine with a rotor with hybrid excitation. International Patent WO2011036135A1.

Grellet, G. and Clerc, G. (1997). *Actionneurs électriques. Principes, modèles, commande*. Editions Eyrolles, Paris.

Hadji-Minaglou, J.-R. and Henneberger, G. (1999). Comparison of different motor types for electric vehicle application. *EPE J.*, 8(3–4), 46–55.

Harris, M.R., Lawrenson, P.J., Stephenson, J.M. (1970). *Per-Unit Systems with Special Reference to Electrical Machines*. Cambridge University Press in association with the Institution of Electrical Engineers, Cambridge.

HBM (2021). T40B : mesure de couple de 50 Nm à 10 kNm [Online]. Available at: https://www.hbm.com/fr/3004/t40b-torque-transducer-with-a-rotational-speed-measuring-system/ [Accessed 9th November 2021].

Henneberger, G., Kuppers, S., Ramesohl, I. (1996). Numerical calculation, simulation and design optimisation of claw pole alternators for automotive application. *IEE Colloq. Mach. Automot. Appl.*, London, 3/1–3/5.

Hlioui, S. (2008). Étude d'une machine synchrone à double excitation. Contribution à la mise en place d'une plate-forme de logiciels en vue d'un dimensionnement optimal. PhD Thesis, Université de Technologie de Belfort-Montbeliard.

Hlioui, S., Amara, Y., Hoang, E., Lecrivain, M., Gabsi, M. (2013). Overview of hybrid excitation synchronous machines technology. *Proc. Int. Conf. Elect. Eng. Softw. Appl.*, (ICEESA), Hammamet, 1–10.

Hlioui, S., Gabsi, M., Ben Ahmed, H., Barakat, G., Amara, Y., Chabour, F.J., Paulides, J.H. (2022). Hybrid excited synchronous machines. *IEEE Trans. Magn.*, 58(2), 8101610.

Hoang, E., Ben Ahmed, A.H., Lucidarme, J. (1997). Switching flux permanent magnet polyphased synchronous machines. *Proc. EPE-97 Conf.*, 3, 903–908.

Hoang, E., Lécrivain, M., Gabsi, M. (2007). A new structure of a switching flux synchronous polyphased machine with hybrid excitation. *Proceedings of the 2007 European Conference on Power Electronics and Applications*, Aalborg.

Hoang, E., Lecrivain, M., Gabsi, M. (2011). Flux-switching dual excitation electrical machine. U.S. Patent US7868506B2.

Hua, H. and Zhu, Z.Q. (2017). Novel partitioned stator hybrid excited switched flux machines. *IEEE Trans. Energy Convers.*, 32(2), 495–504.

Hua, H. and Zhu, Z.Q. (2020). Comparative study of series and parallel hybrid excited machines. *IEEE Trans. Energy Convers.*, 35(3), 1705–1714.

Ionel, D.M., Eastham, J.F., Betzer, T. (1995). Finite element analysis of a novel brushless DC motor with flux barriers. *IEEE Trans. MAG*, 31(6), 3749–3751.

Ionel, D.M., Balchin, M.J., Eastham, J.F., Demeter, E. (1996). Finite element analysis of brushless DC motors for flux weakening operation. *IEEE Trans. Magn.*, 32(5), 5040–5042.

Jahns, T.M. (1987). Flux-weakening regime of an interior permanent-magnet synchronous motor drive. *IEEE Trans. Industry Appl.*, IA-23, 681–689.

Jahns, T.M. (2000). Component rating requirements for wide constant power operation of interior PM synchronous machine drives. *Industry Applications Conference, Conference Record of the 2000 IEEE*, Rome, 3, 1697–1704.

Jahns, T.M., Kliman, G.B., Neumann, T.W. (1986). Interior permanent-magnet synchronous motors for adjustable-speed drives. *IEEE Trans. Ind. Applicat.*, IA-22, 738–747.

Kamiev, K. (2013). Design and testing of armature-reaction compensated permanent magnet synchronous generator for island operation. PhD Thesis, Lappeenranta University of Technology.

Kamiev, K., Nerg, J., Pyrhönen, J., Zaboin, V., Hrabovcová, V., Rafajdus, P. (2012). Hybrid excitation synchronous generators for island operation. *IET Electric Power Applications*, 6(1), 1–11.

Kamiev, K., Pyrhönen, J., Nerg, J., Zaboin, V., Tapia, J. (2013). Modeling and testing of an armature-reaction-compensated (PM) synchronous generator. *IEEE Trans. Energy Convers.*, 28(4), 849–859.

Kamiev, K., Nerg, J., Pyrhönen, J., Zaboin, V., Tapia, J. (2014). Feasibility of an armature-reaction-compensated permanent-magnet synchronous generator in island operation. *IEEE Trans. Ind. Electron.*, 61(9), 5057–5058.

Kron, G. (1930). Generalized theory of electrical machinery. *AIEE Trans.*, 49, 663–683.

Lacroux, G. (1994). *Les actionneurs électriques pour la robotique et les asservissements*, 2nd revised and corrected edition. Tec & Doc, Paris.

Lajoie-Mazenc, M. and Viarouge, P. (1991). Alimentation des machines synchrones. Dossier D3630, Techniques de l'Ingénieur.

Lawali Ali, H. (2021). Étude de structures de générateurs linéaires pour la conversion de l'énergie de la houle. PhD Thesis, Université Le Havre Normandie, Normandie Université.

Lee, C.H. (1963). Vernier motor and its design. *IEEE Trans. Power Apparatus & Sys.*, PAS-82, 343–349.

Leonardi, F., Matsuo, T., Li, Y., Lipo, T.A., MacCleer, P.J. (1996). Design considerations and test results for a doubly salient PM motor with flux control. *Conf. Rec. IEEE IAS Annu. Meeting*, 2, 836–842.

Leonhard, W. (1997). *Control of Electrical Drives*. Springer, Berlin.

Levy, A. (1997). Quelle motorisation pour les véhicules électriques : synchrone à rotor bobiné ou à aimants ? *C-VELEC'97*, 68–74.

Li, X., Shen, F., Yu, S., Xue, Z. (2021). Flux-regulation principle and performance analysis of a novel axial partitioned stator hybrid-excitation flux-switching machine using parallel magnetic circuit. *IEEE Trans. Ind. Electron.*, 68(8), 6560–6573.

Liu, J., Yang, X., Zhang, J., Xue, J., Gao, F., Xiao, C., Gao, Y., Tan, F. (2018). Hybrid excitation motor for new energy automobile. Chinese Patent CN108429421A.

Louis, J.-P. (1999). Claude Bergman. Commande numérique des machines synchrones. Dossier D3644, Techniques de l'Ingénieur.

Louis, J.-P. (2010). *Commandes classiques et avancées des actionneurs synchrones*. Hermès-Lavoisier, Paris.

Louis, J.-P. (2011). *Commandes d'actionneurs électriques synchrones et spéciaux*. Hermès-Lavoisier, Paris.

Luo, X. and Lipo, T.A. (2000). A synchronous/permanent magnet hybrid AC machine. *IEEE Trans. Energy Convers.*, 15(2), 203–210.

Mailfert, A. and Sargos, F.-M. (2004). Machines à réluctance variable (MRV) – Machines polyphasées. Machines excitées. Dossier D3681, Techniques de l'Ingénieur.

Miller, T.J.E. (1989). *Brushless Permanent-Magnet and Reluctance Motor Drives*. Calderon Press, Oxford.

Miyamasu, M. and Akatsu, K. (2011). Efficiency comparison between brushless DC motor and brushless AC motor considering driving method and machine design. *Proc. of the 37th Annual Conference of the IEEE Industrial Electronics Society*, IECON, Melbourne.

Mizuno, T. (1997). Hybrid excitation type permanent magnet synchronous motor. US Patent 5682073.

Mizutani, R., Tatematsu, K., Yamada, E., Matsui, N., Kosaka, T. (2008). Rotating electric motor. International Patent WO2008093865A1.

Mohan, N., Undeland, T.M., Robbins, W.P. (1995). *Power Electronics: Converters, Applications, and Design*, 2nd edition. John Wiley & Sons, Inc., Hoboken.

Morimoto, S., Tong, Y., Takeda, Y., Hirasa, T. (1994). Loss minimization control of permanent magnet synchronous motor drives. *IEEE Trans. Ind. Electron.*, 41(5), 511–517.

Moynot, V., Chabot, F., Lecrivain, M., Gabsi, M., Hlioui, S. (2010). Rotating electric machine with homopolar double excitation. International Patent WO2010052439A2.

Multon, B., Lucidarme, J., Prévond, L. (1995). Analyse des possibilités de fonctionnement en régime de désexcitation des moteurs à aimants permanents. *Journal de Physique III*, 5(5), 623–640.

Nam, K.H. (2019). *AC Motor Control and Electrical Vehicle Applications*, 2nd edition. CRC Press, Taylor and Francis Group, Boca Raton.

Nasr, A. (2017). Nouvelles structures de machines électriques pour la génération embarquée avionique. PhD Thesis, Université Paris-Saclay.

Nasr, A., Hlioui, S., Gabsi, M., Mairie, M., Lalevee, D. (2017). Design optimization of a hybrid-excited flux-switching machine for aircraft-safe DC power generation using a diode bridge rectifier. *IEEE Trans. Ind. Electron.*, 64(12), 9896–9904.

Nedjar, B. (2011). Modélisation basée sur la méthode des réseaux de perméances en vue de l'optimisation de machines synchrones à simple et à double excitation. PhD Thesis, École normale supérieure de Cachan.

Nidec, L.S. (2021). Manuel de l'électromécanique [Online]. Available at: https://www.leroy-somer.com/documentation_pdf/5181_fr.pdf [Accessed 9 November 2021].

Pothi, N., Zhu, Z.Q., Afinowi, I.A.A., Lee, B., Ren, Y. (2015). Control strategy for hybrid-excited switched-flux permanent magnet machines. *IET Elect. Power Appl.*, 9(9), 612–619.

Reutlinger, K. (2008). Electric machine comprising a rotor with hybrid excitation. International Patent WO2008148621A1.

Richard, D. and Dubel, Y. (2007). Valeo StARS technology: A competitive solution for hybridization. *Proceedings of the Power Conversion Conference, PCC'07*, 1601–1605, Nagoya.

RLS (2020). RM44/RM58 rotary magnetic encoder with AM4096 [Online]. Available at: https://www.rls.si/cn_tw/fileuploader/download/download/?d=1&file=custom%2Fupload%2FRM44D06_02.pdf [Accessed 27 July 2022].

Roboam, X., Sareni, B., Andrade, A. (2012). More electricity in the air: Toward optimized electrical networks embedded in more-electrical aircraft. *IEEE Ind. Electron. Mag.*, 6(4), 6–17.

Rosenberg, E. (1940). Securing correct polarity generators. U.S. Patent US2207304.

Rosenberg, H. (1968). Permanent magnet excited electric machine. U.S. Patent US3411027.

Schiferl, R.F. and Lipo, T.A. (1990). Power capability of salient pole permanent magnet synchronous motors in variable speed drive applications. *IEEE Trans. Ind. Appl.*, 26(1), 115–123.

Shafranek, R.J. (1967). Self-excited brushless alternator. U.S. Patent US3346749.

Shi, Y.F., Zhu, Z.Q., Howe, D. (2006). Torque-speed characteristics of interior-magnet machine in brushless AC and DC modes, with particular reference to their flux-weakening performance. *Proc. Int. Power Electronics and Motion Control Conf.*, IPEMC, Shanghai.

Sneyers, B., Novotny, D.W., Lipo, T.A. (1985). Field weakening in buried permanent magnet AC motor drives. *IEEE Trans. Ind. Appl.*, IA-21, 398–407.

Soong, W.L. and Miller, T.J.E. (1994). Field-weakening performance of brushless synchronous AC motor drives. *IEE Proc. Electr. Power Appl.*, 141(6), 331–340.

Soulard, J. (1998). Étude paramétrique des ensembles convertisseur machine à aimants : application à une structure électromagnétique monophasée hybride à aimants permanents et à alimentation électronique. PhD Thesis, Université Paris VI.

Soulard, J. and Multon, B. (1999). Maximum power limits of small single-phase permanent magnet drives. *IEE Proc. Electr. Power Appl.*, 146(5), 457–462.

Spooner, E., Khatab, S.A.W., Nicolaou, N.G. (1989). Hybrid excitation of AC and DC machines. *Proc. 1989, 4th International Conference on Electrical Machines and Drives Conf.*, London, 48–52.

Sulaiman, E., Kosaka, T., Matsui, N. (2011). High power density design of 6-slot–8-pole hybrid excitation flux switching machine for hybrid electric vehicles. *IEEE Trans. Magn.*, 47(10), 4453–4456.

Sun, X. and Zhu, Z.Q. (2020). Investigation of DC winding induced voltage in hybrid-excited switched-flux permanent magnet machine. *IEEE Trans. Ind. Appl.*, 56(4), 3594–3603.

Syverson, D. (1995). Hybrid alternator. U.S. Patent US5397975A.

Takorabet, A. (2008). Dimensionnement d'une machine à double excitation de structure innovante pour une application alternateur automobile : comparaison à des structures classiques. PhD Thesis, Ecole Normale Supérieure de Cachan.

Tapia, J.A., Leonardi, F., Lipo, T.A. (2003). Consequent-pole permanent magnet machine with extended field-weakening capability. *IEEE Trans. Ind. Appl.*, 39(6), 1704–1709.

Terman, F.E. (1926). The circle diagram of a transmission network. *AIEE Trans.*, 45, 1081–1092.

Terman, F.E., Lenzen, T.L., Freedman, C.L., Rogers, K.A. (1930). The general circle diagram of electrical machinery. *A. I. E. E. Journal*, 16–18.

Thomas, A. (2017). *Lipo: Introduction to AC Machine Design*. IEEE Press, Piscataway.

Tiegna, H. (2013). Contribution à la modélisation analytique des machines synchrones à flux axial à aimants permanents à attaque directe en vue de leur dimensionnement. Application aux éoliennes. PhD Thesis, Université du Havre.

Vas, P. (1992). *Electrical Machines and Drives: A Space-Vector Theory Approach*. Clarendon Press, Oxford.

Vido, L. (2004). Étude d'actionneurs électriques à double excitation destinés au transport : dimensionnement de structures synchrones. PhD Thesis, École normale supérieure de Cachan.

Vido, L., Amara, Y., Gabsi, M., Lecrivain, M., Chabot, F. (2005a). Compared performances of homopolar and bipolar hybrid excitation synchronous machines. *Conf. Rec. Industry Appl. Conf., 40th IAS Annu. Meeting*, 3, 1555–1560.

Vido, L., Gabsi, M., Lécrivain, M., Amara, Y., Chabot, F. (2005b). Homopolar and bipolar hybrid excitation synchronous machines. *Proc. IEEE Int. Electric Machines and Drives Conf.*, IEMDC, San Antonio, 1212–1218.

Vido, L., Amara, Y., Gabsi, M. (2011). Machines synchrones à double excitation MSDE. Techniques de l'Ingénieur, D3525 V1.

Wang, G. (2009). A hybrid excitation synchronous generator. International Patent WO2009082875A1.

Wang, Y. and Deng, Z. (2012a). Comparison of hybrid excitation topologies for flux-switching machines. *IEEE Trans. Magn.*, 48(9), 2518–2527.

Wang, Y. and Deng, Z. (2012b). Hybrid excitation topologies and control strategies of stator permanent magnet machines for DC power system. *IEEE Trans. Ind. Electron.*, 59(12), 4601–4616.

Wang, H., Zhao, C., Guo, H. (2010). Hybrid excitation brushless synchronous motor. Chinese Patent CN101752969A.

Wardach, M., Palka, R., Paplicki, P., Prajzendanc, P., Zarebski, T. (2020). Modern hybrid excited electric machines. *Energies*, 13, 5910.

Wu, Z., Zhu, Z.Q., Hua, W., Akehurst, S., Zhu, X., Zhang, W., Hu, J., Li, H., Zhu, J. (2019a). Analysis and suppression of induced voltage pulsation in DC winding of five-phase wound-field switched flux machines. *IEEE Trans. Energy Convers.*, 34(4), 1890–1905.

Wu, Z.Z., Zhu, Z.Q., Wang, C., Mipo, J.-C., Personnaz, S., Farah, P. (2019b). Reduction of open-circuit DC-winding-induced voltage in wound field switched flux machines by skewing. *IEEE Trans. Ind. Electron.*, 66(3), 1715–1726.

Wu, Z., Zhu, Z.Q., Wang, C., Mipo, J.-C., Personnaz, S., Farah, P. (2020). Analysis and reduction of on-load DC winding induced voltage in wound field switched flux machines. *IEEE Trans. Ind. Electron.*, 67(4), 2655–2666.

Xia, G., Zeng, Q., Cai, Y. (2010). Tangential-set magnet double salient hybrid excitation motor. Chinese Patent CN201403037Y.

Xie, S.J. and Xu, Z. (2010). Multitooth magnetic bridge type hybrid excitation magnetic flux switching motor. Chinese Patent CN101834474A.

Yamamura, S. (1986). *AC Motors for High-Performance Applications*. Marcel Dekker, New York.

Yamazaki, K., Nishioka, K., Shima, K., Fukami, T., Shirai, K. (2012). Estimation of assist effects by additional permanent magnets in salient pole synchronous generators. *IEEE Trans. Ind. Electron.*, 59(6), 2515–2523.

Yang, H., Zhu, Z.Q., Lin, H., Zhan, H.L., Hua, H. (2012). Hybrid-excited switched-flux hybrid magnet memory machines. *IEEE Trans. Magn.*, 52(6), 8202215.

Yang, H., Zhu, Z.Q., Lin, H., Chu, W.Q. (2016). Flux adjustable permanent magnet machines: A technology status review. *Chin. J. Elect. Eng.*, 2(2), 14–30.

Yang, H., Li, Y., Lin, H., Zhu, Z.Q., Lyu, S., Wang, H., Fang, S., Huang, Y. (2017). Novel reluctance axis shifted machines with hybrid rotors. *Proc. Energy Convers. Congr. and Expo. (ECCE)*, 2362–2367.

Yu, H., Hu, M., Shi, L., Qin, F. (2009). Hybrid excitation linear synchronous motor using Halbach permanent magnet. Chinese Patent CN101594040A.

Zeng, Z. and Lu, Q. (2018). Investigation of novel partitioned-primary hybrid-excited flux-switching linear machines. *IEEE Trans. Ind. Electron.*, 65(12), 9804–9813.

Zhao, W., Chen, D., Lipo, T.A., Kwon, B.-I. (2015a). Performance improvement of ferrite-assisted synchronous reluctance machines using asymmetrical rotor configurations. *IEEE Trans. Magn.*, 51(11), art. no. 8108504.

Zhao, W., Lipo, T.A., Kwon, B.I. (2015b). Optimal design of a novel asymmetrical rotor structure to obtain torque and efficiency improvement in surface inset pm motors. *IEEE Trans. Magn.*, 51(3), 1–4.

Zhu, Z.Q. and Cai, S. (2019). Hybrid excited permanent magnet machines for electric and hybrid electric vehicles. *CES Transactions on Electrical Machines and Systems*, 3(3), 233–247.

Zhu, Z.Q., Shen, J.X., Howe, D. (2006). Flux-weakening characteristics of trapezoidal BACK-EMF machines in brushless DC and AC Modes. *Proc. Int. Power Electronics and Motion Control Conference*, IPEMC, Shanghai.

Zhu, Z.Q., Chen, J.T., Howe, D., Iwasaki, S., Deodhar, R. (2008). Analysis of a novel multi-tooth flux-switching permanent magnet brushless AC machines for high torque direct drives. *IEEE Trans. Magn.*, 44(11), 4313–4316.

Index

2D structures, 15, 19, 24, 25, 153, 172, 173
3D structures, 19, 21, 23, 25, 153, 155

A, C

applications, 3, 4, 10–12, 28
classification, 12, 13, 19, 65
control
   laws, 31, 32, 35, 36, 40, 41, 51–53, 55, 56, 58–60, 63–65, 70, 72–83, 86–89, 95, 96, 98, 101, 103, 107–111, 118, 121, 128, 130
   strategies, 12, 51, 58, 77, 87, 88, 90, 95, 96, 98, 99, 102–104, 110, 112, 113, 132

E, F

efficiency
   maps, 115, 118, 123–125, 127, 128, 130–132, 135, 143–147, 150, 164, 167–169
   optimization, 113
electric vehicles, 35, 36
electrical machines, 12, 28
excitation flux control, 162
flux weakening, 63, 65, 72

G, H

generator operation, 194
hybrid excited synchronous machines (HESM), 1, 3, 5, 6, 12, 13, 18, 19, 23–25, 27, 28
   parallel (PHESM), 13, 15–21, 23, 24, 26
   series (SHESM), 13–15, 17–19, 24, 25, 27, 29
hybridization
   rate, 2
   ratio optimization, 67, 112, 114, 116, 117, 135, 146, 147, 149

M, N

magnetic circuit, 13, 14, 16, 17
manufacturing/construction, 159, 161, 179, 181, 182, 185
measurements, 162, 164, 167, 187–190, 192
motor operation, 161, 162, 164, 194
no-load operation, 155, 159

O, S

operation under load, 188
speed variation, 113, 124, 125

Other titles from

iSTE

in

Systems and Industrial Engineering – Robotics

2022

BOURRIÈRES Jean-Paul, PINÈDE Nathalie, TRAORÉ Mamadou Kaba, ZACHAREWICZ Grégory
From Logistic Networks to Social Networks: Similarities, Specificities, Modeling, Evaluation

DEMOLY Frédéric, ANDRÉ Jean-Claude
4D Printing 1: Between Disruptive Research and Industrial Applications
4D Printing 2: Between Science and Technology

HAJJI Rafika, JARAR OULIDI Hassane
Building Information Modeling for a Smart and Sustainable Urban Space

KROB Daniel
Model-based Systems Architecting: Using CESAM to Architect Complex Systems (Volume 3 - Systemes of Systems Complexity Set)

LOUIS Gilles
Dynamics of Aircraft Flight

2020

BRON Jean-Yves
System Requirements Engineering

KRYSINSKI TOMASZ, MALBURET FRANÇOIS
Energy and Motorization in the Automotive and Aeronautics Industries

PRINTZ Jacques
System Architecture and Complexity: Contribution of Systems of Systems to Systems Thinking

2019

ANDRÉ Jean-Claude
Industry 4.0: Paradoxes and Conflicts

BENSALAH Mounir, ELOUADI Abdelmajid, MHARZI Hassan
Railway Information Modeling RIM: The Track to Rail Modernization

BLUA Philippe, YALAOU Farouk, AMODEO Lionel, DE BLOCK Michaël, LAPLANCHE David
Hospital Logistics and e-Management: Digital Transition and Revolution

BRIFFAUT Jean-Pierre
From Complexity in the Natural Sciences to Complexity in Operations Management Systems
(Systems of Systems Complexity Set – Volume 1)

BUDINGER Marc, HAZYUK Ion, COÏC Clément
Multi-Physics Modeling of Technological Systems

FLAUS Jean-Marie
Cybersecurity of Industrial Systems

JAULIN Luc
Mobile Robotics – Second Edition Revised and Updated

KUMAR Kaushik, DAVIM Paulo J.
Optimization for Engineering Problems

TRIGEASSOU Jean-Claude, MAAMRI Nezha
Analysis, Modeling and Stability of Fractional Order Differential Systems 1: The Infinite State Approach
Analysis, Modeling and Stability of Fractional Order Differential Systems 2: The Infinite State Approach

VANDERHAEGEN Frédéric, MAAOUI Choubeila, SALLAK Mohamed, BERDJAG Denis
Automation Challenges of Socio-technical Systems

2018

BERRAH Lamia, CLIVILLÉ Vincent, FOULLOY Laurent
Industrial Objectives and Industrial Performance: Concepts and Fuzzy Handling

GONZALEZ-FELIU Jesus
Sustainable Urban Logistics: Planning and Evaluation

GROUS Ammar
Applied Mechanical Design

LEROY Alain
Production Availability and Reliability: Use in the Oil and Gas Industry

MARÉ Jean-Charles
Aerospace Actuators 3: European Commercial Aircraft and Tiltrotor Aircraft

MAXA Jean-Aimé, BEN MAHMOUD Mohamed Slim, LARRIEU Nicolas
Model-driven Development for Embedded Software: Application to Communications for Drone Swarm

MBIHI Jean
Analog Automation and Digital Feedback Control Techniques
Advanced Techniques and Technology of Computer-Aided Feedback Control

MORANA Joëlle
Logistics

SIMON Christophe, WEBER Philippe, SALLAK Mohamed
Data Uncertainty and Important Measures
(Systems Dependability Assessment Set – Volume 3)

TANIGUCHI Eiichi, THOMPSON Russell G.
City Logistics 1: New Opportunities and Challenges
City Logistics 2: Modeling and Planning Initiatives
City Logistics 3: Towards Sustainable and Liveable Cities

ZELM Martin, JAEKEL Frank-Walter, DOUMEINGTS Guy, WOLLSCHLAEGER Martin
Enterprise Interoperability: Smart Services and Business Impact of Enterprise Interoperability

2017

ANDRÉ Jean-Claude
From Additive Manufacturing to 3D/4D Printing 1: From Concepts to Achievements
From Additive Manufacturing to 3D/4D Printing 2: Current Techniques, Improvements and their Limitations
From Additive Manufacturing to 3D/4D Printing 3: Breakthrough Innovations: Programmable Material, 4D Printing and Bio-printing

ARCHIMÈDE Bernard, VALLESPIR Bruno
Enterprise Interoperability: INTEROP-PGSO Vision

CAMMAN Christelle, FIORE Claude, LIVOLSI Laurent, QUERRO Pascal
Supply Chain Management and Business Performance: The VASC Model

FEYEL Philippe
Robust Control, Optimization with Metaheuristics

MARÉ Jean-Charles
Aerospace Actuators 2: Signal-by-Wire and Power-by-Wire

POPESCU Dumitru, AMIRA Gharbi, STEFANOIU Dan, BORNE Pierre
Process Control Design for Industrial Applications

YANNIS George, COHEN Simon
Traffic Safety (Research for Innovative Transports Set - Volume 4)

2015

AUBRY Jean-François, BRINZEI Nicolae
Systems Dependability Assessment: Modeling with Graphs and Finite State Automata

BOULANGER Jean-Louis
CENELEC 50128 and IEC 62279 Standards

BRIFFAUT Jean-Pierre
E-Enabled Operations Management

MISSIKOFF Michele, CANDUCCI Massimo, MAIDEN Neil
Enterprise Innovation

2014

CHETTO Maryline
Real-time Systems Scheduling
Volume 1 – Fundamentals
Volume 2 – Focuses

DAVIM J. Paulo
Machinability of Advanced Materials

ESTAMPE Dominique
Supply Chain Performance and Evaluation Models

FAVRE Bernard
Introduction to Sustainable Transports

GAUTHIER Michaël, ANDREFF Nicolas, DOMBRE Etienne
Intracorporeal Robotics: From Milliscale to Nanoscale

MICOUIN Patrice
Model Based Systems Engineering: Fundamentals and Methods

RÉVEILLAC Jean-Michel
Modeling and Simulation of Logistics Flows 1: Theory and Fundamentals
Modeling and Simulation of Logistics Flows 2: Dashboards, Traffic Planning and Management
Modeling and Simulation of Logistics Flows 3: Discrete and Continuous Flows in 2D/3D

2016

ANDRÉ Michel, SAMARAS Zissis
Energy and Environment
(Research for Innovative Transports Set - Volume 1)

AUBRY Jean-François, BRINZEI Nicolae, MAZOUNI Mohammed-Habib
Systems Dependability Assessment: Benefits of Petri Net Models (Systems Dependability Assessment Set - Volume 1)

BLANQUART Corinne, CLAUSEN Uwe, JACOB Bernard
Towards Innovative Freight and Logistics (Research for Innovative Transports Set - Volume 2)

COHEN Simon, YANNIS George
Traffic Management (Research for Innovative Transports Set - Volume 3)

MARÉ Jean-Charles
Aerospace Actuators 1: Needs, Reliability and Hydraulic Power Solutions

REZG Nidhal, HAJEJ Zied, BOSCHIAN-CAMPANER Valerio
Production and Maintenance Optimization Problems: Logistic Constraints and Leasing Warranty Services

TORRENTI Jean-Michel, LA TORRE Francesca
Materials and Infrastructures 1 (Research for Innovative Transports Set - Volume 5A)
Materials and Infrastructures 2 (Research for Innovative Transports Set - Volume 5B)

WEBER Philippe, SIMON Christophe
Benefits of Bayesian Network Models
(Systems Dependability Assessment Set – Volume 2)

MILLOT Patrick
Designing Human–Machine Cooperation Systems

NI Zhenjiang, PACORET Céline, BENOSMAN Ryad, RÉGNIER Stéphane
Haptic Feedback Teleoperation of Optical Tweezers

OUSTALOUP Alain
Diversity and Non-integer Differentiation for System Dynamics

REZG Nidhal, DELLAGI Sofien, KHATAD Abdelhakim
Joint Optimization of Maintenance and Production Policies

STEFANOIU Dan, BORNE Pierre, POPESCU Dumitru, FILIP Florin Gh., EL KAMEL Abdelkader
Optimization in Engineering Sciences: Metaheuristics, Stochastic Methods and Decision Support

2013

ALAZARD Daniel
Reverse Engineering in Control Design

ARIOUI Hichem, NEHAOUA Lamri
Driving Simulation

CHADLI Mohammed, COPPIER Hervé
Command-control for Real-time Systems

DAAFOUZ Jamal, TARBOURIECH Sophie, SIGALOTTI Mario
Hybrid Systems with Constraints

FEYEL Philippe
Loop-shaping Robust Control

FLAUS Jean-Marie
Risk Analysis: Socio-technical and Industrial Systems

FRIBOURG Laurent, SOULAT Romain
Control of Switching Systems by Invariance Analysis: Application to Power Electronics

GROSSARD Mathieu, REGNIER Stéphane, CHAILLET Nicolas
Flexible Robotics: Applications to Multiscale Manipulations

GRUNN Emmanuel, PHAM Anh Tuan
Modeling of Complex Systems: Application to Aeronautical Dynamics

HABIB Maki K., DAVIM J. Paulo
Interdisciplinary Mechatronics: Engineering Science and Research Development

HAMMADI Slim, KSOURI Mekki
Multimodal Transport Systems

JARBOUI Bassem, SIARRY Patrick, TEGHEM Jacques
Metaheuristics for Production Scheduling

KIRILLOV Oleg N., PELINOVSKY Dmitry E.
Nonlinear Physical Systems

LE Vu Tuan Hieu, STOICA Cristina, ALAMO Teodoro, CAMACHO Eduardo F., DUMUR Didier
Zonotopes: From Guaranteed State-estimation to Control

MACHADO Carolina, DAVIM J. Paulo
Management and Engineering Innovation

MORANA Joëlle
Sustainable Supply Chain Management

SANDOU Guillaume
Metaheuristic Optimization for the Design of Automatic Control Laws

STOICAN Florin, OLARU Sorin
Set-theoretic Fault Detection in Multisensor Systems

2012

AÏT-KADI Daoud, CHOUINARD Marc, MARCOTTE Suzanne, RIOPEL Diane
Sustainable Reverse Logistics Network: Engineering and Management

BORNE Pierre, POPESCU Dumitru, FILIP Florin G., STEFANOIU Dan
Optimization in Engineering Sciences: Exact Methods

CHADLI Mohammed, BORNE Pierre
Multiple Models Approach in Automation: Takagi-Sugeno Fuzzy Systems

DAVIM J. Paulo
Lasers in Manufacturing

DECLERCK Philippe
Discrete Event Systems in Dioid Algebra and Conventional Algebra

DOUMIATI Moustapha, CHARARA Ali, VICTORINO Alessandro, LECHNER Daniel
Vehicle Dynamics Estimation using Kalman Filtering: Experimental Validation

GUERRERO José A, LOZANO Rogelio
Flight Formation Control

HAMMADI Slim, KSOURI Mekki
Advanced Mobility and Transport Engineering

MAILLARD Pierre
Competitive Quality Strategies

MATTA Nada, VANDENBOOMGAERDE Yves, ARLAT Jean
Supervision and Safety of Complex Systems

POLER Raul *et al.*
Intelligent Non-hierarchical Manufacturing Networks

TROCCAZ Jocelyne
Medical Robotics

YALAOUI Alice, CHEHADE Hicham, YALAOUI Farouk, AMODEO Lionel
Optimization of Logistics

ZELM Martin *et al.*
Enterprise Interoperability –I-EASA12 Proceedings

2011

CANTOT Pascal, LUZEAUX Dominique
Simulation and Modeling of Systems of Systems

DAVIM J. Paulo
Mechatronics

DAVIM J. Paulo
Wood Machining

GROUS Ammar
Applied Metrology for Manufacturing Engineering

KOLSKI Christophe
Human–Computer Interactions in Transport

LUZEAUX Dominique, RUAULT Jean-René, WIPPLER Jean-Luc
Complex Systems and Systems of Systems Engineering

ZELM Martin, *et al.*
Enterprise Interoperability: IWEI2011 Proceedings

2010

BOTTA-GENOULAZ Valérie, CAMPAGNE Jean-Pierre, LLERENA Daniel, PELLEGRIN Claude
Supply Chain Performance / Collaboration, Alignement and Coordination

BOURLÈS Henri, GODFREY K.C. Kwan
Linear Systems

BOURRIÈRES Jean-Paul
Proceedings of CEISIE'09

CHAILLET Nicolas, REGNIER Stéphane
Microrobotics for Micromanipulation

DAVIM J. Paulo
Sustainable Manufacturing

GIORDANO Max, MATHIEU Luc, VILLENEUVE François
Product Life-Cycle Management / Geometric Variations

LOZANO Rogelio
Unmanned Aerial Vehicles / Embedded Control

LUZEAUX Dominique, RUAULT Jean-René
Systems of Systems

VILLENEUVE François, MATHIEU Luc
Geometric Tolerancing of Products

2009

DIAZ Michel
Petri Nets / Fundamental Models, Verification and Applications

OZEL Tugrul, DAVIM J. Paulo
Intelligent Machining

PITRAT Jacques
Artificial Beings

2008

ARTIGUES Christian, DEMASSEY Sophie, NERON Emmanuel
Resources–Constrained Project Scheduling

BILLAUT Jean-Charles, MOUKRIM Aziz, SANLAVILLE Eric
Flexibility and Robustness in Scheduling

DOCHAIN Denis
Bioprocess Control

LOPEZ Pierre, ROUBELLAT François
Production Scheduling

THIERRY Caroline, THOMAS André, BEL Gérard
Supply Chain Simulation and Management

2007

DE LARMINAT Philippe
Analysis and Control of Linear Systems

DOMBRE Etienne, KHALIL Wisama
Robot Manipulators

LAMNABHI Françoise *et al.*
Taming Heterogeneity and Complexity of Embedded Control

LIMNIOS Nikolaos
Fault Trees

2006

FRENCH COLLEGE OF METROLOGY
Metrology in Industry

NAJIM Kaddour
Control of Continuous Linear Systems